SpringerBriefs in Optimization

SpringerBriefs in Optimization showcases algorithmic and theoretical techniques, case studies, and applications within the broad-based field of optimization. Manuscripts related to the ever-growing applications of optimization in applied mathematics, engineering, medicine, economics, and other applied sciences are encouraged.

More information about this series at http://www.springer.com/series/8918

Wen Xu • Weili Wu

Optimal Social Influence

 Springer

Wen Xu
Department of Mathematics
and Computer Science
Texas Woman's University
Denton, TX, USA

Weili Wu
Department of Computer Science
University of Texas, Dallas
Richardson, TX, USA

ISSN 2190-8354   ISSN 2191-575X   (electronic)
SpringerBriefs in Optimization
ISBN 978-3-030-37774-8   ISBN 978-3-030-37775-5   (eBook)
https://doi.org/10.1007/978-3-030-37775-5

This Springer imprint is published by the registered company Springer Nature Switzerland AG.
The registered company address is: Gewerbestrasse 11, 6330 Cham, Switzerland

Preface

Social computing is an emerging research field that studies the computational aspects of social networks. Social influence specifically studies the behaviors of individuals and the influence diffusion among them in a social network, which involves interdisciplinary fields such as mathematics, sociology, psychology, computer science, business, and public safety.

This book is a collection of chapters illustrating optimal social influence. It focuses on recent and practical applications, models, algorithms, and open topics for future research. The main topics cover characteristics of social networks, modeling of social influence propagation, popular research problems in social influence analysis such as influence maximization, rumor blocking, rumor source detection, and multiple social influence competing. The authors present some of the latest social computing research and illustrate how to understand and manipulate social influence for knowledge discovery by applying various data mining techniques in the real-world scenarios.

There are a total of five chapters, which form a complete series of optimal social influence, and meanwhile, are self-contained to provide the greatest reading flexibility. Chapter 1 introduces basic concepts on how to characterize both social networks and social influence. Chapter 2 illustrates diffusion of influence in social networks. Chapter 3 presents a practical application: rumor source detection in social networks. In Chap. 4, we show current research on how to block rumor propagation efficiently in social networks. In the last chapter, we focus on modeling multiple social influence competing in a network.

The target audience of this book includes but not limited to researchers, scholars, graduate students, and developers who are interested in social influence research. The book also aims to serve as a textbook for graduate courses such as computational social networks. We gratefully acknowledge the support, encouragement, and patience of Professor Ding-Zhu Du.

Denton, TX, USA Wen Xu
Dallas, TX, USA Weili Wu
October 2019

Contents

Chapter 1
Introduction of Social Influence Analysis

1.1 Introduction

With the emergence and rapid proliferation of social applications and media, such as instant messaging (e.g., WhatsApp, Viber, WeChat, Snapchat, Line, Facebook Messenger, and Google Hangouts), sharing sites (e.g., Flickr, YouTube, and Yelp), blogs (e.g., WordPress and LiveJournal), wikis (e.g., Wikipedia and PBWiki), microblogs (e.g., Twitter and Weibo), social networks (e.g., Facebook), and collaboration networks (e.g., DBLP), there is little doubt that social influence is becoming a prevalent, complex, and subtle force that governs the dynamics of all social networks. Therefore, social influence study has started to attract intense attention due to many important applications.

Social influence occurs when one's opinions, emotions, or behaviors are affected by others, intentionally or unintentionally. Generally there are two alternative models of influence resulting from behavior in response to referrals: normative influence where recipient behavior is based on interpreting the information provided by the influencer as an implied expectation to conform, or informational influence where recipient behavior is based on a personal evaluation of the information provided by the influencer. For instance, a friend recommends strongly a new restaurant that you have never heard of before and then you tried it. This is called informational social influence since you get new information from others. Another example, all your friends are using social applications such as Twitter or Facebook. Although you are not a fan of social applications, you still use it to keep connections with all the friends. This is called normative social influence, which is to conform to the positive expectations of others.

In the context of normative influence, the mechanisms influencing actions are identification and compliance. Recipient behavior is driven by the desire to maintain the relationship with the influencer and/or be associated with a referent group by fitting in in order to evoke a favorable response from the group. The recipient's willingness to conform is stronger when recipient behavior is observable to the influencer and to others in the social network. In contrast, the mechanism underlying informational influence is internalization, where behavior occurs only when such action is congruent with the recipient's value system and by a personal evaluation of the benefits. Such compliance involves independent information processing by recipients with the goal of maximizing outcomes for themselves.

Overall, behaviors linked to normative influence are driven by recipients' desire to comply, and direct benefits from the action for recipients are often a secondary consideration. In contrast, behavior linked to informational influence arises from an evaluation of the direct benefits to the recipient. While behavior linked to normative influence is often discontinued when recipient action is not observable or salient to the influencer or the group, behavior in response to informational influence is usually sustained and incorporated into habitual actions of respondents over time.

Three Degrees of Social Influence Three degrees of influence is a theory in the realm of social networks, proposed by Nicholas A. Christakis and James H. Fowler in 2007 [43]. Christakis and Fowler found that social networks have great influence on individuals' behavior. But social influence does not end with the people to whom a person is directly tied. We influence our friends who in their turn influence their friends, meaning that our actions can influence people we have never met. They posit that "everything we do or say tends to ripple through our network, having an impact on our friends (one degree), our friends' friends (two degrees), and even our friends' friends' friends (three degrees). Our influence gradually dissipates and ceases to have a noticeable effect on people beyond three degrees of separation". This argument is basically that peer effects need not stop at one degree, and that, if we can affect our friends, then we can (in many cases) affect our friends' friends, and so on.

Studies by Christakis and Fowler suggested that a variety of attributes—like obesity [43], smoking [44], and happiness [68]—rather than being individualistic, are casually correlated by contagion mechanisms that transmit these behaviors over long distances within social networks. The three degrees of influence property has also been observed in criminal networks [203]. This applies to many aspects of life, from public health to economics. For instance, it might be preferable to immunize individuals located in network's center more than peripheral individuals. Or, it might be much more effective to motivate clusters or people to avoid criminal behavior than to act upon individuals or than to punish each criminal separately.

If people are connected to everyone by six degrees of separation[134] and influence those up to three degrees [68], then people can reach halfway to anyone in the world. The strength of social influence depends on many factors such as the strength of relationships between people in the networks, the network distance between users, temporal effects, characteristics of networks and individuals in the network. Also, nodes (users, entities) are influenced by others for various reasons. For example, the colleagues have strong influence on one's work, while the friends have strong influence on one's daily life. How to prove theoretically or empirically that social influence does exist? How to quantify the strength of those social influences? What kind of factors decides the strength of social influence in real large networks?

In this chapter, we focus on computational aspect of social influence analysis and describe the measures and algorithms related to it. In Sect. 1.2, we introduce fundamental concepts about modeling social networks, which are deeply related to the importance or influence of nodes or edges in the networks. In Sects. 1.3 and 1.4, we aim at qualitatively or quantitatively measuring the influence levels of nodes and edges in the network.

1.2   Characteristics of Social Networks

A clear understanding of social influence in social networks cannot be achieved without a clear understanding of characteristics of social networks. Therefore, in this section, we give an overview about social networks identifying its significance and characteristics. Although levels of analysis are not necessarily mutually exclusive, there are three general levels into which networks may fall: micro-level, meso-level, and macro-level.

At the micro-level, social network analysis typically emphasizes on social relationship between social actors instead of the attributes of social actors. It studies a small group of individuals in a particular social context such as dyads and triads, where a dyad is a social relationship between two individuals and a triad is social relationships among three individuals. For an overview of basic social network analysis techniques, we refer the reader to the book by Wasserman and Faust [199]. There are many interesting properties about social networks that have been studied by sociologists. In 1967, Milgram [134] shows that the average path length between two Americans is six hops. By way of introduction, from any other person in the world, so that a chain of "a friend of a friend" statements can be made to connect any two people in a maximum of six steps, which is called "six degree of separation" theory. In 2003, Columbia University conducted an analogous experiment on social connectedness among Internet email users, which kind of confirmed Milgram's theory. In 2013 [121], Lou et al. predict reciprocity and triadic closure in social networks by proposing a triad factor graph (TriFG) model, which incorporates social theories into a semi-supervised learning method. On a large Twitter network, they show that with the proposed factor graph model it is possible to accurately infer 90% of reciprocal relationships in a dynamic network.

At the meso-level, research begins with a population size that falls between the micro- and macro-levels. As a type of complex network, social network has some well-known theoretical properties such as power-law degree distributions, the small-world property, and the scale-free property. Power-law distribution means that the probability that a node has degree k is proportional to $k^{-\gamma}$, for large k and $\gamma > 1$. The parameter γ is called the power-law coefficient. Researchers have shown that many real-world networks are power-law networks, including power grids [153], neural networks [27], Internet topologies [61], the Web [15, 106], and social networks [1]. Scale-free networks are a class of power-law networks where the high-degree nodes tend to be connected to other high degree nodes. Scale-free graphs are discussed in detail by Li et al. [117], and they propose a metric to measure the scale-freeness of graphs. Small-world networks mean a class of network in which most nodes are not neighbors of one another, but most nodes can be reached from every other by a small number of hops or steps, which generally have a small diameter and exhibit high clustering. Studies have shown that the Web [4, 29], scientific collaboration on research papers [143], film actors [6], and general social networks [1] have small-world properties. Pool and Kochen [155] provide an analysis of the small-world effect. The influential paper by Granovetter [77] argues that a social network can be partitioned into strong and weak ties, and that the strong ties are tightly clustered. Kleinberg [99, 100] proposes a model to explain the small-world phenomenon in offline social networks and also examines navigability in these networks.

A prominent study of the Web link structure [29] shows that the Web has a bow-tie shape, consisting of a single large strongly connected component (SCC), and other groups of nodes that can either reach the SCC or can be reached from the SCC. Online social networks have a similar large component, but that its relative size is much larger than that of the Web's SCC. Faloutsos et al. [61] show that the degree distribution of the Internet follows a power-law. Siganos et al. demonstrate that the high-level structure of the Internet resembles a jellyfish [178]. Kleinberg [102] demonstrates that high-degree pages in the Web can be identified by their function as either hubs (containing useful references on a subject) or authorities (containing relevant information on a subject). Kleinberg also presents an algorithm [101] for inferring which pages function as hubs and which as authorities. The well-known PageRank algorithm [150] uses the Web structure to determine pages that contain authoritative information.

As online social networks are gaining popularity, sociologists and computer scientists are beginning to investigate their properties. Adamic et al. [1] study an early online social network at Stanford University, and find that the network exhibits small-world behavior, as well as significant local clustering. Liben-Nowell et al. [118] find a strong correlation between friendship and geographic location in social networks by using data from LiveJournal. Kumar et al. [107] analyze two online social networks and discover that both possess a large strongly connected component. Girvan and Newman observe that users in online social networks tend to form tightly knit groups, which is also called communities [72]. Backstrom et al. [12] examine snapshots of group membership in LiveJournal and present models for the growth of user groups over time.

At the macro-level, there are many recent research works that focus on large-scale social networks and big data analysis. Social networks especially online social networks have become good platforms for generating and collecting large volume of user data. Driven by real-world applications in e-commerce, market intelligence, e-government, healthcare, and security [42], initialized by national funding agencies, managing and mining big data have shown to be a challenging yet very compelling task. In 2014, Wu et al. [204] present a HACE theorem that characterizes a conceptual view of the big data processing framework, which includes three levels: data, model, and system. With big data technologies, we will hopefully be able to provide most relevant and most accurate social sensing feedback to better understand our society at real-time.

To summarize, these are the following important structural properties of social networks:

- The degree distributions in social networks follow a power-law, and the power-law coefficients for both in-degree and outdegree are similar. Nodes with high in-degree also tend to have high outdegree.
- Social networks appear to be composed of a large number of highly connected clusters consisting of relatively low-degree nodes. These clusters connect to each other via a relatively small number of high-degree nodes. As a consequence, the clustering coefficient is inversely proportional to node degree.
- Social networks are small-world networks containing large, densely connected cores. Overall, the network is held together by about 10% of the nodes with highest degree. As a result, path lengths are short, but almost all shortest paths of sufficient length traverse the highly connected core.
- Online social networks, which contain rich customer opinion and behavioral information, have become highly scalable ecommerce platforms and product recommendation systems. Most recent research focuses on social networks as large-volume, heterogeneous, autonomous sources with distributed and decentralized control, and seeks to explore complex and evolving relationships among data.

The measurements indicate that online social networks have a high degree of reciprocity, contain tight cores that consist of high-degree nodes, and provide good platforms for collecting various types of data. These findings are likely applicable to many different applications, especially on information dissemination, search, and influence inference.

1.3 Measuring Social Influence

In this book, a social network is modeled as a graph $G = \{V, E\}$, where V is the set of nodes representing individuals, and E is the set of edges corresponding to social relationships. We next classify metrics of social influence into three types, edge based metrics, node based metrics, and user action based metrics. For edge based metrics, social influence is a directional effect from node u to node v and is related to the edge strength from u to v. In terms of node based metrics, some nodes can have intrinsically higher influence than others due to the network structure. There are also works capturing social influence according to the user actions and historical data.

1.3.1 Edge Based Metrics

Edge measures study the simple influence-related processes and interactions between individuals.

Tie Strength According to Granovetter's seminal work [77], the tie strength between two nodes depends on the overlap of their neighborhoods. In particular, the more common neighbors that a pair of nodes have, the stronger the tie between them. If the overlap of neighborhoods between two nodes, say A and B, is large, we consider A and B have a strong tie. Otherwise, they are considered to have a weak tie. We formally define the strength $S(A, B)$ in terms of their Jaccard coefficient.

$$S(A, B) = \frac{n_A \bigcap n_B}{n_A \bigcup n_B} \qquad (1.1)$$

where n_A and n_B indicate the neighborhoods of A and B, respectively. Sometimes, the tie strength is defined under a different name called embeddedness.The embeddedness of an edge is high if two nodes incident on the edge have a high overlap of neighborhoods. When two individuals are connected by an embedded edge, it makes it easier for them to trust one another, because it is easier to find out dishonest behavior through mutual friends [79]. On the other end, when embeddedness is zero, two end nodes have no mutual friends. Therefore, it is riskier for them to trust each other because there are no mutual friends for behavioral verification.

A corollary from the tie strength is the hypothesis of triadic closure. This relates to the nature of the ties between sets of three nodes A, B, and C. If strong ties connect A to B and A to C, then B and C are likely to be connected by a strong tie as well. Conversely, if $A - B$ and $A - C$ are weak ties, B and C are less likely to have a strong tie. Triadic closure is measured by the clustering coefficient of the network[200]. The clustering coefficient of a node A is defined as the probability that two randomly selected friends of A are friends with each other. In other words, it is the fraction of pairs of friends of A that are linked to one another. This is naturally related to the problem of triangle counting problem in a network. Let n_Δ be the number of triangles in the network and $|E|$ be the number of edges. The clustering coefficient is formally defined as follows:

$$C = \frac{6n_\Delta}{|E|} \qquad (1.2)$$

The naive way of counting the number of triangles n_Δ is expensive. An interesting connection between n? and the eigenvalues of the network was discovered by Tsourakakis [193]. This work shows that n_Δ is approximately equal to the third-moment of the eigenvalues (or $\sum \lambda_i^3$, where λ_i is the ith eigenvalue). Given the skewed distribution of eigenvalues, the triangle counts can be approximated by computing the third-moment of only a small number of the top eigenvalues. This also provides an efficient way for computing the clustering coefficient.

Weak Ties When the overlap of the neighborhoods of A and B is small, the connection $A - B$ is considered to be a weak tie. When there is no overlap, the connection $A - B$ is a local bridge. In the extreme case, the removal of $A - B$ may result in the disconnection of the connected component containing A and B. In such a case, the connection $A - B$ may be considered a global bridge. It may be argued that in real networks, global bridges occur rarely as compared to local bridges. However, the effect of local and global bridges is quite similar.

Edge Betweenness Another important measure is the edge betweenness, which measures the total amount of flow across the edge. Here, we assume that the information flow between A and B is evenly distributed on the shortest paths between A and B. Freeman [70] first articulated the concept of betweenness in the context of sociology. One application of edge betweenness is that of graph partitioning. The idea is to gradually remove edges of high betweenness scores to turn the network into a hierarchy of disconnected components. These disconnected components will be the clusters of nodes in the network. More detailed studies on clustering methods are presented in the work by Girvan and Newman [72].

1.3.2 Node Based Metrics

One node based metric measuring the importance of a node in a network is centrality. Centrality has attracted a lot of attention as a tool for studying social networks. A node with high centrality score is usually considered more highly influential than other nodes in the network.

Many centrality measures have been proposed based on the specific definition of influence. The main principle to categorize the centrality measures is the type of random walk computation involved. In particular, the centrality measures can be grouped into two categories: radial and medial measures [24]. Radial measures assess random walks that start or end from a given node. On the other hand, medial measures assess random walks that pass through a given node. The radial measures are further separated into volume measures and length measures based on the type of random walks. Volume measures fix the length of walks and find the volume (or number) of the walks limited by the length. Length measures fix the volume of the target nodes and find the length of walks to reach the target volume. Next we introduce some popular centrality measures based on these categories.

Degree Centrality The first group of the centrality measures is that of the radial and volume based measures. The simplest and most popular measure in this category is that of degree centrality. Let A be the adjacency matrix of a network, and $deg(i)$ be the degree of node i. The degree centrality c_i^{DEG} of node i is defined to be the degree of the node:

$$c_i^{DEG} = deg(i) \tag{1.3}$$

One way of interpreting the degree centrality is that it counts the number of paths of length 1 that starts from a node. A natural generalization from this perspective is the K-path centrality which is the number of paths of length at most k that start from a node.

Another class of measures are based on the diffusion behavior in the network. The Katz centrality [92] counts the number of walks starting from a node, while penalizing longer walks. More formally, the Katz centrality c_i^{KATZ} of node i is defined as follows:

$$c_i^{KATZ} = e_i^T (\sum_{j=1}^{\infty}(\beta A)^j)\mathbf{1} \tag{1.4}$$

Here, e_i is a column vector whose ith element is 1, and all other elements are 0. The value of β is a positive penalty constant between 0 and 1.

Closeness Centrality The second group of the centrality measures is that of the radial and length based measures. Unlike the volume based measures, the length based measures count the length of the walks. The most popular centrality measure in this group is the Freeman's closeness centrality [70]. It measures the centrality by computing the average of the shortest distances to all other nodes. Then, the closeness centrality c_i^{CLO} of node i is defined as follows:

$$c_i^{CLO} = e_i^T S\mathbf{1} \tag{1.5}$$

Here S be the matrix whose (i, j)th element contains the length of the shortest path from node i to j and $\mathbf{1}$ is the all one vector.

Node Betweenness As is the case for edges of high betweenness, nodes of high betweenness occupy critical positions in the network structure, and are therefore able to play critical roles. This is often enabled by a large amount of flow, which is carried by nodes which occupy a position at the interface of tightly knit groups. Such nodes are considered to have high betweenness. The concept of betweenness is related to nodes that span structural holes in a social network. We will discuss more on this point slightly later.

Another popular group of the centrality measures is that of medial measures. It is called "medial" since all the walks passing through a node are considered. The most well-known centrality in this group is the Freeman's betweenness centrality [69]. It measures how much a given node lies in the shortest paths of other nodes. The betweenness centrality c_i^{BET} of node i is defined as follows:

$$c_i^{BET} = \sum_{j,k} \frac{b_{jik}}{b_{jk}} \tag{1.6}$$

Here b_{jk} is the number of shortest paths from node j to k, and b_{jik} be the number of shortest paths from node j to k that pass through node i.

The naive algorithm for computing the betweenness involves all-pair shortest paths. This requires $T(n^3)$ time and $T(n2)$ storage. Brandes [28] designed a faster algorithm with the use of n single-source-shortest-path algorithms. This requires $O(n + m)$ space and runs in $O(nm)$ and $O(nm + n^2 log n)$ time, where n is the number of nodes and m is the number of edges.

Newman [144] proposed an alternative betweenness centrality measure based on random walks on the graph. The main idea is that instead of considering shortest paths, it considers all possible walks and computes the betweenness from these different walks. Then, the Newman's betweenness centrality c_i^{NBE} of node i is defined as follows:

$$c_i^{NBE} = \sum_{j \neq i \neq k} R_{jk}^{(i)} \tag{1.7}$$

Here $R^{(i)}$ be the matrix whose (j, k)th element $R_{jk}^{(i)}$ contains the probability of a random walk from j to k, which contains i as an intermediate node.

Structural Holes In a network, we call a node a structural hole if it is connected to multiple local bridges. A canonical example is that a person's success within a company or organization often depends on their access to local bridges [31]. By removing such a person, an "empty space" will occur in the network. This is referred to as a structural hole. The person who serves as a structural hole can interconnect information originating from multiple noninteracting parties. Therefore, this person is structurally important to the communication between the actor representing a structural hole and the organization that may not be aligned. For the organization, accelerating the information flow between groups could be beneficial, which requires building of bridges. However, this building of bridges would come at the expense of structural hole's latent power of regulating information flow at the boundaries of these groups.

1.3.3   User Behavior Based Metrics

The aforementioned metrics concentrate on the structure of the network rather than the behavior of nodes and their interactions. However, influence is usually reflected in changes in social action patterns (user behavior) in the social network. Some researchers consider other aspects of analysis such as analysis of users' behavior in a network. Recent work [75, 205] has studied the problem of learning the influence degree from historical user actions, while some other work [169, 184] investigates how social actions evolve in the context of the network, and how such actions are affected by social influence factors. Before introducing these methods, we will first define the time-varying attribute-augmented networks with user actions:

Definition 1.1 Time-varying attribute-action augmented network: The time-varying attribute-action augmented network is denoted as $Gt = (V_t, E_t, X_t, Y_t)$, where V_t is the set of users and E_t is the set of links between users at time t, X_t represents the attribute matrix of all users in the network at time t, and Y_t represents the set of actions of all users at time t.

For all actions, they define a set of action tuples as $Y = (v, y, t)$, where $v \in V_t$, $t \in 1, \ldots, T$, and $y \in Y_t$.

Goyal et al. [75] study the problem of learning the influence probabilities from a historic log of user actions. They present the concept of user influential probability and action influential probability. The assumption is that if user v_i performs an action y at time t and later $(t' > t)$ his friend v_j also performs the action, then there is an influence from v_i on v_j. The goal of learning influence probabilities is to find a (static of dynamic) model to best capture the user influence and action influence in the network. They give a general user influence probability and action influence probability definitions as follows:

User influence probability

$$inf(v_i) = \frac{|y|\exists v, \Delta t : prop(a, v_i, v_j, \Delta t) \bigwedge 0 \leq \Delta t|}{Y_{v_i}} \tag{1.8}$$

Action influence probability

$$inf(y) = \frac{|v_i|\exists v_j, \Delta t : prop(a, v_j, v_i, \Delta t) \bigwedge 0 \leq \Delta t|}{number of users performing y} \tag{1.9}$$

where $\Delta t = t_j - t_i$ represents the difference between the time when user v_j performing the action and the time when user v_i performing the action, given $e_{ij} = 1$; $prop(a, v_i, v_j, \Delta t)$ represents the action propagation score.

Goyal et al. [75] propose three methods to approximate the action propagation $prop(a, v_i, v_j, \Delta t)$: (1) static model based on Bernoulli distribution, Jaccard index and partial credits, (2) continuous time (CT) model, and (3) discrete time (DT) model. The model can be learned with a two-stage algorithm. Finally, the learned influence probabilities have been applied to action prediction and the experiments show that the continuous time (CT) model can achieve a better performance than other models on the Flickr social network with the action of "joining a group."

Meeyoung Cha et al. [36] studied the measurement of influence in Twitter based on the following three metrics:

1. In-degree: is the number of followers of a user. In-degree represents the popularity of a user in a social network.
2. Retweets: which is defined as the number of times that followers of a specific node pass-along a posting from a tweeter. Retweeting causes propagation of a posting or news in a network. This metric is important as it shows how an advertisement can propagate across the network using influential users.

3. Mentions: this means the number of times that the name of a user is mentioned in his followers' postings. This metric has been observed to follow a power-law distribution.

Meeyoung Cha et al. [36] used Spearman's rank correlation coefficient for comparing users' influence. The research compared the three measures mentioned above to analyze topics of the most influential people in Twitter according to the aforementioned analysis and retweets.

Afrasiabi and Benyoucef [159] studied influence as a combination of link strength and incoming and outgoing clustering value defined for each node in the network. The link strength is measured according to the volume of interactions among users while the clustering value is measured by the closeness of a node to highly interconnected communities. They filter the spam and inactive nodes according to their activities and their interaction with other users.

Dynamic graph analysis has been studied by Khrabrov and Cybenko [95] in which the number of daily mentions for each user is considered as an indicator for computing different ranks such as PageRank, drank, and starrank for influence analysis of each node in a network. For example, starrank considers user importance with respect to his or her neighborhoods. These researchers used several primitive indices in combination, such as Contiguous Longest Increasing Subsequences (CLIS) and GrowFall for analysis of influence ranks during a period of time. These indices show how the influence rank of a user changes with time. Khrabrov and Cybenko also analyzed the rate of increase in the number of mentions for influencing users in a network for consecutive days.

About this topic, readers may further refer to [82, 182], which present excellent surveys on how to measure social influence. Here we included their methods and further improve them on the completeness.

1.4   Identifying Social Influence

A central problem for social influence is to understand the interplay between similarity and social ties. A lot of research has tried to identify influence and correlation in social networks from many different aspects: social similarity and influence [7, 48]; marketing through social influence [53, 161], influence maximization [93]; social influence model and practice through conformity, compliance and obedience [46, 58], and social influence in virtual worlds [57].

1.4.1   Homophily

Homophily [112] is one of the most fundamental principle that governs the structure of social networks—which means that, an actor in the social network tends to be similar to his connected neighbors or "friends." This is a natural result, because the friends or neighbors of a given actor in the social network are not a random sample of the underlying population. The neighbors of a given actor in the social network are often similar to that actor along many different dimensions including racial and ethnic dimensions, age, their occupations, and their interests and beliefs. Singla et al. [179] have conducted a large-scale experiment of homophily on real social networks, which includes data from user interactions in the MSN Messenger network and a subset of Microsoft Web search data collected in the summer of 2006. They observe that the similarities between two friends are significantly larger than a random selected pairwise sample, especially in attributes such as age, location, and query category. This experiment confirms the existence of homophily at a global scale in large online social networks.

The phenomenon of homophily can originate from many different mechanisms:

Social Influence This indicates that people tend to follow the behaviors of their friends. The social influence effect leads people to adopt behaviors exhibited by their neighbors.

Selection This indicates that people tend to create relationships with other people who are already similar to them.

Confounding Variables Other unknown variables exist, which may cause friends to behave similarly with one another.

These three factors are often intertwined in real social networks, and the overall effect is to provide a strong support for the homophily phenomenon. Intuitively, the effects of selection and social influence lead to different applications in mining social network data. In particular, recommendation systems are based on the selection/social similarity, while viral marketing [161] is based on social influence. To model these different factors, several models have been proposed [48, 88].

Generative Models for Selection and Influence Holme and Newman [88] proposed a generative model to balance the effects of selection and influence. The idea is to initially place the M edges of the network uniformly at random between vertex pairs and also assign opinions to vertices uniformly at random. With this initialization, an influence and selection based dynamic is simulated. Each step of the simulation either moves an edge to lie between two individuals whose opinions agree (selection process), or we change the opinion of an individual to agree with one of their neighbors (influence process). The results of their simulation confirmed that the selection tend to generate a large number of small clusters, while social influence will generate large coherent clusters. Thus, this interesting model suggests that these two factors both support clusters in the network, though the nature of such clusters is quite different.

Every vertex in the Holme-Newman model [88] at a given time can only have one opinion. This may be an oversimplification of real social networks. To address this limitation, Crandall et al. [48] introduced multidimensional opinion vectors to better model complex social networks. In particular, they assumed that there is a set of m possible activities in the social network. Each node v at time t has an m-dimensional vector $v(t)$, where the ith coordinate of $v(t)$ represents the extent to which person v is engaging in activity i. They use cosine similarity to compute the similarity between two people. Similar to the Holme-Newman model, Crandall et al. also propose a more comprehensive generative model which samples a person's activities based on their own history, their neighbors' history, and a background distribution. Crandall's model is arguably more powerful, but also requires more parameters. Therefore more data is required in order to learn the parameters. Finally, they applied their model and conducted a predictive modeling study on Wikipedia and live journal datasets. The benefit of the proposed similarity model is still inconclusive.

Quantifying Influence and Selection Subsequent to the work in [48], Scripps et al. [169] proposed the formal computational definitions of selection and influence. We formally define selection and influence as follows:

$$Selection = \frac{p(a_{ij}^t = 1 | a_{ij}^{t-1} = 0, < x_i^{t-1}, x_j^{t-1} >> \epsilon)}{p(a_{ij}^t = 1 | a_{ij}^{t-1} = 0)} \qquad (1.10)$$

Here, the denominator is the conditional probability that an unlinked pair will become linked and the numerator is the same probability for unlinked pairs whose similarity exceeds the threshold ϵ. Values greater than one indicate the presence of selection.

$$Influence = \frac{p(< x_i^t, x_j^t >>< x_i^{t-1}, x_j^{t-1} > | a_{ij}^{t-1} = 0, a_{ij}^t = 1)}{p(< x_i^t, x_j^t >>< x_i^{t-1}, x_j^{t-1} > | a_{ij}^{t-1} = 0)} \qquad (1.11)$$

Here, the numerator is the conditional probability that similarity increases from time $t-1$ to t between two nodes that became linked at time t and the denominator is the probability that the similarity increases from time $t-1$ to t between two nodes that were not linked at time $t-1$. As with selection, values greater than one indicate the presence of influence.

Based on this definition, Scripps et al. [169] present a matrix alignment framework by incorporating the temporal information to learn the weight of different attributes for establishing relationships between users. This can be done by optimizing (minimizing) the following objective function:

$$min W \sum_{t-1}^{T} \| A^t - X^{t-1} W X^{(t-1)} T \|_F^2 \qquad (1.12)$$

where the diagonal elements of W correspond to the vector of weights of attributes and $\| \cdot \|_F$ denotes the Frobenius norm. Solving the objective function (Eq. (1.12)) is equivalent to the problem of finding the weights of different attributes associated with users. A distortion distance function is used to measure the degree of influence and selection.

The above method can be used to analyze influence and selection. However, several theories in sociology [77] show that the effect of the social influence from different angles (topics) may be different. To differentiate the influence from different angles (topics), Tang et al. [186] propose a topical factor graph (TFG) model to formalize the topic-level social influence analysis into a unified graphical model and present topical affinity propagation (TAP) for model learning.

1.4.2   Existential Test

Identifying social influence in networks is critical to understanding how behaviors spread. Ugandera et al. [196] study who influenced us and it shows that our behavior is influenced by the "structural diversity" (the number of connected components in our ego network) instead of the number of our friends. Anagnostopoulos et al. [7] try to differentiate social influence from homophily or confounding variables by proposing the shuffle test and edge-reversal test. The idea of shuffle test is that if social influence does not play a role, even though an agent's probability of activation could depend on her friends, the timing of such an activation should be independent of the timing of other agents. Therefore, the data distribution and characteristics will not change even if the exact time of occurrence is shuffled around. The idea of edge-reversal test is that other forms of social correlation (than social influence) are only based on the fact that two friends often share common characteristics or are affected by the same external variables and are independent of which of these two individuals has named the other as a friend. Thus, reversing the edges will not change our estimate of the social correlation significantly. On the other hand, social influence spreads in the direction specified by the edges of the graph, and hence reversing the edges should intuitively change the estimate of the correlation. Anagnostopoulos et al. [7] test their models using tagging data from Flickr and validate social influence as a source of correlation between the actions of individuals with social ties.

The proposed tests in [7] assume a static network, which is true in many real social networks. LaFond and Neville [65] propose a different randomization test with the use of a relational autoregression model. More specifically, they propose to model the social network as a time-evolving graph $G_t = (V, E_t)$ where V is the set of all nodes and E_t is the set of all edges at time t. Besides G_t, the nodes have some attribute at time t denoted by X_t. The main idea is that selection and social influence can be differentiated through the autocorrelation between X_t and G_t. On the one hand, the selection process can be represented as a causal relationship from X_{t-1} to G_t, which means the node attributes at time $t-1$, i.e., X_{t-1}, determines the social network at G_t. On the other hand, the social influence can be represented as the causal relation from G_{t-1} to X_t, which means the social network at time t, i.e., G_t, determines the node attributes at time t, i.e., X_t.

Aral et al. [11] propose a diffusion model for differentiating selection and social influence. In particular, their intuition is that although the diffusion patterns created by peer influence-driven contagions and homophilous diffusion are similar, the effects are likely to result in significantly different dynamics. Influence-driven contagions are self-reinforcing and display rapid, exponential, and less predictable diffusion as they evolve, whereas selection-driven diffusion processes are governed by the distributions of characteristics over nodes. In [11], they develop a matched sample estimation framework to distinguish influence and homophily effects in dynamic networks.

Social Influence in Healthcare   Christakis and Fowler studied the effect of social influence on health related issues including alcohol consumption [163], obesity [43], smoking [44], trouble sleep [131], loneliness [32], and happiness [68]. In these studies, they use longitudinal data covering roughly 12,000 people and correlate health status and social network structure over a 32-year period. They found clusters of nodes with similar health status in the network. In other words, people tend to be more similar in health status to their friends than in a random graph. The main focus of all these studies is to explain why homophily of health status is present. The analysis in Christakis and Fowler argues that, even accounting for effects of selection and confounding variables, there is significant evidence for social influence as well. The evidence suggests that health status can be influenced by the health status of the neighbors. For example, their obesity study [43] suggests that obesity may exhibit some amount of "contagion" in the social network. Although people do not necessarily catch it as the way one catches the flu, it can spread through the underlying social network via the mechanism of social influence. Similar observations of their study on alcohol consumption [163] discover that clusters of drinkers and abstainers were present in the network at all time points, and the clusters extended to three degrees of separation through the social network. These clusters were not only due to selective formation of social ties among drinkers but also seem to reflect social influence. Changes in the alcohol consumption behavior of a person's social network had a statistically significant effect on that person's subsequent alcohol consumption behavior. The behaviors of immediate neighbors and co-workers were not significantly associated with a person's drinking behavior, but the behavior of relatives and friends was.

Social Influence in Political Mobilization   Human behavior is thought to spread through face-to-face social networks, but it is difficult to identify social influence effects in observational studies, and it is unknown whether online social networks operate in the same way. In 2012, Bond et al. [22] conduct a randomized controlled trial of political mobilization messages delivered to 61 million Facebook users during the 2010 US congressional elections.

The experimental objects (users) were divided into three groups: social message group in which users were shown with message that indicates one's friends who have made the votes; informational message group, in which users were shown with message that indicates how many of his friends have voted; the last group, control group, in which users did not receive any message. Three groups of users are compared according to their votes.

The results show that the messages directly influenced political self-expression, information seeking, and real-world voting behavior of millions of people. Furthermore, the messages not only influenced the users who received them but also the users' friends and friends of friends. The effect of social transmission on real-world voting was greater than the direct effect of the messages themselves, and nearly all the transmission occurred between "close friends" who were more likely to have a face-to-face relationship. These results suggest that strong ties are instrumental for spreading both online and real-world behavior in human social networks.

1.4.3 Influence-Related Topics

Influence and Interaction Besides the attribute and user actions, influence can be also reflected by the interactions between users. Typically, online communities contain ancillary interaction information about users. For example, a Facebook user has a Wall page, where her friends can post messages. Based on the messages posted on the Wall, one can infer which friends are close and which are acquaintances only. Similarly, one can use follower and following members on Twitter to infer the strength of a relationship.

Xiang et al. [205] propose a latent variable model to infer relationship strength based on profile similarity and interaction activity, with the goal of automatically distinguishing strong relationships from weak ones. The model attempts to represent the intrinsic causality of social interactions via statistical dependencies. It distinguishes interaction activity from user profile data and integrates two types of information by considering the relationship strength to be the hidden effect of user profile similarities, as well as the hidden cause of the interactions between users.

The input to the problem can be considered an attribute-augmented network $G = (V, E, X)$ with interaction information $m_{ij} \subset M$ between users, where m_{ij} is a set of different interactions between users v_i and v_j. The model also uses continuous latent variable z, but for each link rather than action. The latent variable can be further treated as the strength of the social influence. There are some methods aiming to model social influence using a link analysis method. The basic idea is similar to the concept of random walks. Java et al. [89] employ such a method to model the influence in online social networks.

While converting a blog network into an influence graph, a link from u to v indicates that u is influenced by v. The edges in the influence graph are the reverse of the blog graph to indicate this influence. Multiple edges indicate stronger influence and are weighted higher. In the influence graph, the direction of edges is opposite as the blog graph. And the influence weight can be calculated by $W_{u,v} = C_{u,v}/d_v$.

Based on the influence graph, they proposed several typical applications, such as spam detection and node selection. The classical PageRank and HITS algorithms can be also employed here.

Influential Verse Susceptible In 2012, Aral et al. [10] present a method that uses in vivo randomized experimentation to identify influence and susceptibility in networks while avoiding the biases inherent in traditional estimates of social contagion. Estimation in a representative sample of 1.3 million Facebook users showed that younger users are more susceptible to influence than older users, men are more influential than women, women influence men more than they influence other women, and married individuals are the least susceptible to influence in the decision to adopt the product offered. Analysis of influence and susceptibility together with network structure revealed that influential individuals are less susceptible to influence than noninfluential individuals and that they cluster in the network while susceptible individuals do not, which suggests that influential people with influential friends may be instrumental in the spread of this product in the network.

Influence and Friendship Drift Sarkar et al. [167] study the problem of friendships drifting over time. They explore two aspects of social network modeling by the use of a latent space model. First, they generalize a static model of relationships into a dynamic model that accounts for friendships drifting over time. Second, they show how to make it tractable to learn such models from data, even as the number of entities n gets large. The generalized model associates each entity with a point in p-dimensional Euclidean latent space. The points can move as time progresses but large moves in latent space are improbable. Observed links between entities are more likely if the entities are close in latent space. They show how to make such a model tractable (subquadratic in the number of entities) by the use of the following characteristics: (a) appropriate kernel functions for similarity in latent space; (b) the use of low dimensional KD-trees; (c) a new efficient dynamic adaptation of multidimensional scaling for a first pass of approximate projection of entities into latent space; and (d) an efficient conjugate gradient update rule for non-linear local optimization in which amortized time per entity during an update is $O(logn)$. They use both synthetic and real data on up to 11,000 entities which indicate near-linear scaling in computation time and improved performance over four alternative approaches. We also illustrate the system operating on 12 years of NIPS co-authorship data.

Influence and Autocorrelation Autocorrelation refers to correlation between values of the same variable (e.g., action or attribute) associated with linked nodes (users) [142]. More formally, autocorrelation in social networks, and in particular for influence analysis, can be defined with respect to a set of linked users $e_{ij} = 1$, $e_{ij} \in E$, and an attribute matrix X associated with these uses, as the correlation between the values of X on these instance pairs.

Neville et al. provide an overview of research on autocorrelation in a number of fields with an emphasis on implications for relational learning, and outline a number of challenges and opportunities for model learning and inference [142]. Social phenomena such as social influence, diffusion processes, and the principle of homophily give rise to autocorrelated observations as well, through their influence on social interactions that govern the data generation process.

Another related topic is referred to as collective behavior in social networks. Essentially, collective behavior modeling is to understand the behavior correlation in the social network. For this purpose, much work has been done. For example, Tang and Liu [185] aim to predict collective behaviors in social media. In particular, they try to answer the question: given information about some individuals, how can we infer the behavior of unobserved individuals in the same network?

They attempt to utilize the behavior correlation presented in a social network to predict the collective behavior in social media. The input of their problem is the same as Definition 1.1. They propose a framework called SocDim [184], which is composed of two steps, which are those of social dimension extraction and discriminative learning, respectively. In the instantiation of the framework SocDim, modularity maximization is adopted to extract social dimensions. There are several concerns about the scalability of SocDim:

(a) The social dimensions extracted according to modularity maximization are dense.
(b) The modularity maximization requires the computation of the top eigenvectors of a modularity matrix which will become a daunting task when the network scales to millions of node.
(c) Networks in social media tend to evolve which entails efficient update of the model for collective behavior prediction.

Influence and Grouping Behavior Grouping behavior, e.g., user's participation behavior into a forum, is an important action in the social network. The point of influence and grouping behavior is to study how different factors influence the dynamics of grouping behaviors.

Shi et al. investigated the user participation behavior in diverse online forums [176]. In that paper, they are mainly focused on three central questions:

(a) What are the factors in online forums that potentially influence people's behavior in joining communities and what is the corresponding impact?
(b) What are the relationships between these factors, i.e., which ones are more effective in predicting the user joining behavior, and which ones carry supplementary information?
(c) What are the similarities and differences of user grouping behavior in forums of different types (such as news forums versus technology forums)?

In order to answer the first question, they analyze four features that can usually be obtained from a forum dataset:

1. Friends of Reply Relationship. Use this feature to describe how users are influenced by the numbers of neighbors with whom they have ever had any reply relationship.
2. Community Sizes. Use community size as the measurement to quantify the "popularity" of information.

3. Average Ratings of Top Posts. Aside from the popularity of information, we are also interested in how the authority or interestingness of information impacts user behavior.
4. Similarities of Users. This is the only feature with dependency: if two users are "similar" in a certain way, what is the correlation of the sets of communities they join?

Their first discovery is that, despite the relative randomness, the diffusion curve of influence from users of reply relationships has very similar diffusion patterns. However, the reasons that people are linked together are very different. They also investigate the influence of the features associated with communities, which include the size of communities and the authority or the interestingness of the information in the communities. They find that their corresponding information diffusion curves show some strong regularities of user joining behavior as well, and these curves are very different from those of reply relationships. Furthermore, they analyze the effects of similarity of users on the communities they join and find two users who communicate more frequently or have more common friends are more likely to be in the same set of communities.

In order to answer the second question, they construct a bipartite graph, whose two sets of nodes are users and communities, to encompass all the features and their relationships in this problem. Based on the bipartite graph, they build a bipartite Markov random field (BiMRF) model to quantitatively evaluate how much each feature affects the grouping behavior in online forums, as well as their relationships with each other. BiMRF is a Markov random graph with edges and two-stars as its configuration and incorporates the node-level features we have described as in a social selection model. The most significant advantage of using the BiMRF model is that it can explicitly incorporate the dependency between different users' joining behavior, i.e., how a user's joining behavior is affected by her friends' joining behavior. The results of this quantitative analysis show that different features have different effectiveness of prediction in news forums versus technology forums.

Backstrom et al. [13] also explore a large corpus of thriving online communities. These groups vary widely in size, moderation, and privacy and cover an equally diverse set of subject matter. They present a number of descriptive statistics of these groups. Using metadata from groups, members, and individual messages, they identify users who post and are replied to frequently by multiple group members. They classify these high-engagement users based on the longevity of their engagements. Their results show that users who will go on to become long-lived, highly engaged user experience significantly better treatment than other users from the moment they join the group, well before there is an opportunity for them to develop a long-standing relationship with members of the group. They also present a simple model explaining long-term heavy engagement as a combination of user-dependent and group-dependent factors. Using this model as an analytical tool, they show that properties of the user alone are sufficient to explain 95% of all memberships, but introducing a small amount of group-specific information dramatically improves our ability to model users belonging to multiple groups.

1.5 Summary

Social influence analysis aims at qualitatively and quantitatively measuring the influence of one person on others. As social networking becomes more prevalent in the activities of millions of people on a day-to-day basis, both research study and practical applications on social influence will continue to grow. Furthermore, the size of the networks on which the underlying applications need to be used also continues to grow over time. Therefore, effective and efficient social influence methods are in high demand.

In this chapter, we focus on the computational aspects of social influence analysis and describe different methods and algorithms for calculating and identifying social influence-related measures. First, we cover the basic statistical measure of social influence such as centrality, closeness, betweenness, and user behavior based methods; second, we present the models of identifying social influence. These covers the fundamental concepts on influence.

In the future, an important and challenging research area is to develop efficient, effective, and quantifiable social influence mechanisms to enable various applications in social networks and social media. This area lies in the intersection of computer science, business intelligence, cyber security, sociology, etc. In particular, scalable and parallel data mining algorithms and scalable database and web technology have been changing how sociologists approach this problem. In the next chapter, we will focus on discussing diffusion of social influence and influence maximization in viral marketing.

Chapter 2
Diffusion of Information

In this chapter, we outline the techniques used in optimizing or facilitating information diffusion in social networks. We identify two problem definitions through which a broad survey of techniques in recent research is provided. Namely, we explore the problems of maximizing the spread of influence and minimizing the spread of misinformation in social networks. As different as these problems are in terms of the motivation behind them, they both rely on sub-problems that are very similar. Through our study of these two problems, we delve into more detail about the sub-problems: Sect. 2.2 model formation, Sect. 2.3 problem optimization, Sect. 2.4 large-scale data analysis, and Sect. 2.5 research trends.

2.1 Introduction

Diffusion of influence refers to circumstances where a point of view or behavior is widely spread in specific structures of propagation channels [35]. A diffusion can be associated with topological properties, such as scale, range, and temporal properties. This concept has been widely researched in the field of epidemiology, sociology, and marketing.

In early time, biology and epidemiology have conducted in-depth study on diffusion of virus within the group [8], and two classical models: SIS and SIR are proposed. In sociology and marketing area, research on diffusion focuses on the problems of innovation diffusion. In the early twentieth century, Schumpeter et al. [168] created innovative theory. Then the BASS model [3] opened up new research directions for this research area and derived a series of related models. Westerman et al. [202] studied the effect of system generated reports of connectedness on credibility and showed that there are curvilinear effects for the number of followers exist, such that having too many or too few connections results in lower judgments of expertise and trustworthiness. Lopez-Pintado et al. [120] studied the product

W. Xu, W. Wu, *Optimal Social Influence*, SpringerBriefs in Optimization,
https://doi.org/10.1007/978-3-030-37775-5_2

diffusion in complex social networks. He considered the mutual influence among individuals on the micro-level into the propagation equation based on mean-field theory and found out that innovation diffusion in complex networks has a threshold which is closely related to the degree distribution and propagation functions of the network.

Understanding, capturing, and being able to predict influence diffusion can be helpful for several areas such as viral marketing, cyber security, and Web search. For instance, if we consider the case of marketing, it may be useful to know which are the features that control the process of diffusing information when it is created to, e.g., better advertise a product or to better protect it against attacks on the network. The marketing may also benefit from information such as how many initial users to start with in a marketing campaign (budget optimization), how much time to wait between actions, etc. In the case of security, criminal investigators generally need to understand the information flow between, e.g., members of a given community to extract hints regarding possible guilt or innocence of a person or a group of persons. This is clearly an observation phase where the user wants to understand the route that information took and possible links. Finally, as Web search evolves, if we consider the case of subscriptions to feeds, a propagation prediction model may be useful for the user to, e.g., subscribe to the most interesting topic according to its expected growth (in addition to his interests). This reflects a more active usage of the diffusion prediction.

2.2 Model Formation

Central to optimization problems relating to information diffusion is the problem of identifying the right diffusion model. Therefore, we provide a survey of available models and address the following questions: What are the necessary and sufficient parameters of an accurate model? How can we validate the use of a specific model? How can one obtain data about the parameters? Given the intricacy of human interactions, finding the right diffusion model is still an open problem, even in the presence of the large datasets available today. In this section, we give an overview of the most common propagation models, including epidemic models [78, 85], the "Bass" model [16] for product adoption, and basic diffusion models such as independent cascade (IC) and linear threshold (LT) [56]. The goal is also to learn about fundamental properties of such processes in a variety of settings.

2.2.1 Epidemic Models

Infectious agents have had decisive influences on the history of mankind. Fourteenth century Black Death has taken lives of about a third of Europe's population at the time. The first major epidemic in the USA was yellow fever epidemic in

Philadelphia in 1793, in which 5000 people died out of a population of 50,000. This epidemic has had a major impact on the life and politics of the country. Thucydides describes the Plague of Athens (430–428 BC): 1050 of 4000 soldiers on an expedition died of a disease. Thucydides gives a detailed account of symptoms: some so horrendous that the last one—amnesia—seems a blessing. An interesting feature of this account is that there is no mention of person-to-person contagion, which we now suspect with most new diseases. It was not until the nineteenth century that the person-to-person contagion on was beginning to be discussed. In this book, we will mostly be interested in modeling infectious diseases, where the major means of disease spread comes from the person-to-person interaction.

The practical use of epidemic models must rely heavily on the realism put into the models. This does not mean that a reasonable model can include all possible effects, but rather incorporate the mechanisms in the simplest possible fashion so as to maintain major components that influence disease propagation. Great care should be taken before epidemic models are used for prediction of real phenomena. However, even simple models should, and often do, pose important questions about the underlying mechanisms of infection spread and possible means of control of the disease or epidemic.

We begin with classical papers by Kermack and McKendrick (1927, 1932, and 1933). These papers have had a major influence on the development of mathematical models for disease spread and are still relevant in many epidemic situations. The first of these papers laid out a foundation for modeling infections which, after recovery, confer complete immunity (or in case of lethal diseases—death). The population is taken to be constant—no births or deaths other than from the disease are possible—consistent with the course of an epidemic being short compared with the life time of an individual. If a group of infected individuals is introduced into a large population, a basic problem is to describe the spread of the infection within the population as a function of time. In the course of time the epidemic may come to an end. One of the most important questions in epidemiology is to ascertain whether this occurs only when all of the initially susceptible individuals have contracted the disease or if some interplay of infectivity, recovery, and mortality factors may result in epidemic "die out" with many susceptibles still present in the unaffected population.

Mathematical modeling of infectious diseases is a tool to investigate the mechanisms for outbreak and spread of diseases and to predict the future course in order to control an epidemic. Generally there are several types of epidemic models.

First, stochastic models. The epidemic process has random nature. Stochastic models are used to estimate the probabilistic quantities for the outcome events, such as the probability distribution of extinction time, the probability distribution of final epidemic size, the associate mean, and so on.

Second, deterministic compartmental models. The transition rate from one class (compartment) to the other one is characterized by derivative mathematically. If we assume that the population size is differentiable with respect to time, in the limiting of large population, the time evolution of behavior of each subgroup can be approximated by the deterministic dynamics.

In the category of deterministic compartmental models, there are two classical classes: SIR and SIS. In SIS and SIR epidemic models, individuals in the population are classified according to disease status, either susceptible, infectious, or immune. Healthy ("S" = susceptible) nodes become sick ("I" = infected) stochastically from their infected neighbors with a probability. Alternatively, a sick node becomes healthy ("R" = removed) and open to re-infection with a probability. These two parameters are also referred to as the birth rate and death rate of the virus.

The tipping point, or epidemic threshold, of an SIS epidemic model is the condition under which an infection will die out exponentially quickly irrespective of initial infection, as opposed to spreading out, causing and epidemic. For a survey on SIS and numerous other epidemic models, please refer to Hethcote [85].

2.2.2   Product Adoption Model

The well-known first purchase diffusion models in marketing are those of Bass [16], Fourt and Woodlock [67], and Mansfield [129]. These early models attempted to describe the penetration and saturation aspects of the product diffusion process.

The main impetus underlying diffusion research in marketing is the Bass model. Subsuming the models proposed by Fourt and Woodlock [67] and Mansfield [129], the Bass model assumes that potential adopters of an innovation are influenced by two means of communication—mass media and word of mouth. In its development, it further assumes that the adopters of an innovation comprise two groups. One group is influenced only by the mass media communication (external influence) and the other group is influenced only by the word-of-mouth communication (internal influence). Bass termed the first group "Innovators" and the second group "Imitators." Unlike the Bass model, the model proposed by Fourt and Woodlock [67] assumes that the diffusion process is driven primarily by the mass media communication or the external influence. Similarly, the model proposed by Mansfield [129] assumes this process is driven by word of mouth.

Figures 2.1 and 2.2 are plots of conceptual and analytical structure underlying the Bass model. As noted in Fig. 2.1, the Bass model conceptually assumes that "Innovators" or buyers who adopt exclusively because of the mass media communication or the external influence are present at any stage of the diffusion process. Figure 2.2 shows the analytical structure underlying the Bass model. As depicted, the noncumulative adopter distribution peaks at time $T*$, which is the point of inflection of the S-shaped cumulative adoption curve. Furthermore, the adopter distribution assumes that an initial pm (a constant) level of adopters buy the product at the beginning of the diffusion process. Once initiated, the adoption process is symmetric with respect to time around the peak time T^* up to $2T^*$. That is, the shape of the adoption curve from time T^* to $2T^*$ is the mirror image of the shape of the adoption curve from the beginning of the diffusion fusion process up to time T^*. In general, the Bass model is a popular model appeared at an early stage for product adoption. For more information, please refer to [128].

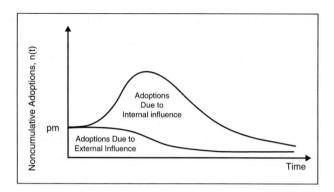

Fig. 2.1 Adoptions due to external and internal influences in the bass model

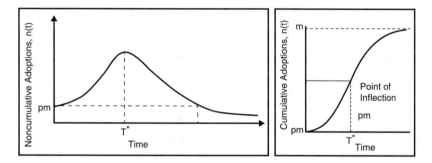

Fig. 2.2 Analytical structure of the bass model

2.2.3 Diffusion Models

Diffusion is the process by which information passes from neighbor to neighbor [137]. Real-world examples include viral marketing, innovation of technologies, and infection propagation. Diffusion models are the framework on which diffusion occurs.

Definition 2.1 A diffusion model is a graph $G = V, E$ along with a collection of activation functions $F = (f_v)_{v \in V}$, where f_v is a $\{\emptyset, \{v\}\}$ valued function on $2^{|V|}$.

The output of a function f_v is a random variable based on the activation function.

Vertices on this graph are usually individuals and the activation function models the influence individuals exert on others. The activation function usually depends only on the neighbors of v, denoted $N(v)$. This means that $f_v(S) = f_v(N(v) \cap S)$.

Definition 2.2 Diffusion is the process on a diffusion model M, $S = (S_t)_{t=0}^{n-1}$ started at $S \subseteq V$:

1. set $S_0 = S$
2. for $t > 1$ set $S_t = f(S_{t-1}) = def \bigcup_{v \in V} f_v(S_{t-1})$

The set of nodes activated at the end of diffusion is denoted as $\sigma(S) = \bigcup_{t=0}^{n}(S_t)$.

Diffusion occurs in time steps t. At each time step, all previously activated nodes remain activated and individuals are either activated or deactivated based on the activation functions. Diffusion can run on a fixed number of time steps or indefinitely. Diffusion is said to have stopped when the set of activated nodes in time step t_k is the same as the set in time step t_{k+n} for all $n \geq 1$.

One class of diffusion models, namely threshold model, adds an influence threshold to each individual, which, when overcome, triggers the individual to be activated. There is a cumulative effect of these models, as it takes a critical number of influential neighbors to activate an individual.

2.2.3.1   General Threshold Model

This model was defined by Kempe et al. [94] and Mossel and Roch [136].

Definition 2.3 The general threshold model is a diffusion model with

1. A set of threshold values $(\theta_v)_{v \in V}$, where θ_v is in the range $[0, 1]$.
2. Node v being activated if $f_v(S) \geq \theta_v$, where S is the set of neighbors of v.

The activation function on the general threshold model depends on the activated neighbors of v. There is an assumption of monotonicity on this model made to reflect that adding active neighbors to a node increases likelihood of the node being activated.

Definition 2.4 A function $f : 2^V \rightarrow R$ is monotone if $f(S) \leq f(T)$ for all $S \subseteq T \subseteq V$.

This property captures that activating more nodes will always have an increasing effect on the nodes that will be activated at a future time.

2.2.3.2   Linear Threshold Model

The linear threshold model is a specialized form of general threshold models. The linear threshold model, LT model in short, is more often used in marketing research.

Definition 2.5 The linear threshold model is a diffusion model with all of the properties of the general threshold model with

1. A set of weights $(p(u, v))_{(u,v) \in E}$ with the property $\Sigma_{u \in N(v)} p(u, v) \leq 1$.
2. Activation function of the form $f_v(S) = \sum_{u \in N(v)} p(v, u)$ with $f(\emptyset) = 0$.

Cascade models of diffusion give each individual the ability to influence their neighbors as soon as they are activated. This is opposed to the threshold models that rely on a cumulative effect. This model has the property that the more nodes that have attempted to influence a node, the less likely the node is to be activated.

Here we give a definition of a specialized cascade model, namely the independent cascade model, IC model in short.

Definition 2.6 The **independent cascade model** is a diffusion model with the following properties:

1. Each arc (u, v) has associated the probability $p(u, v)$ of u influencing v.
2. Time unfolds in discrete steps.
3. At time t, nodes that became active at $t - 1$ try to active their inactive neighbors and succeed according to $p(u, v)$.

Note that the probability of a node u influencing a node v is independent of the set of nodes S that has attempted to influence v.

There is an assumption of monotonicity on this model made to reflect that adding active neighbors to a node increases likelihood of the node being activated.

2.2.3.3   History-Sensitive Cascade Model

The history-sensitive cascade model, designed by Foster and Potter, is essentially a reformat of the linear threshold model and is not a different diffusion model itself. In their research into the spread of influence, Foster and Potter propose the idea that the probability of a node being activated increases the longer the node is in contact with other activated nodes. Since at every time step more neighbors can be added, while the combined influence never goes down, the probability that any node is activated increases with each new neighbor added. This reflects the monotonic property of the linear threshold model.

Foster and Potter studied the exact effects of diffusion over time on the probability that any node would be activated at time step k. They studied this effect on tree-structure graphs and also on general graphs and proposed algorithms for determining these probabilities. To attain the probability of a node being activated at any given time step, a Markov chain model is used.

Definition 2.7 A Markov chain is a sequence of random variables X_1, X_2, X_3, with the property that $Pr(X_{n+1} = x | X_1 = x_1, X_2 = x_2, \ldots, X_n = x_n)$.

A Markov chain is a collection of states with transitions between states such that the probability of transitioning to any state from any other state depends only on the current state. Foster and Potter use a Markov chain model that encode sets of active nodes in binary strings and then create a transition matrix that maps the probability of transitioning from any set of activated nodes to any other set. By iterating over this transition matrix, it is possible to find the exact probability of any node being activated at any time step for any arbitrary graph.

2.2.3.4 Cascade Models

Cascade models of diffusion give each individual the ability to influence their neighbors as soon as they are activated. This is opposed to the threshold models that rely on a cumulative effect. This model has the property that the more nodes that have attempted to influence a node, the less likely the node is to be activated.

2.2.3.5 General Cascade Model

This model was designed by Kempe et al. [94] as a general form of the cascade model. This model has the property that the more nodes that have attempted to influence a node, the less likely the node is to be activated.

Definition 2.8 The general cascade model is a diffusion model with the following properties:

1. nodes are live at time t if they were activated in time $t - 1$.
2. a collection of probability functions $P = p_v, v \in V$ where p_v is a $[0, 1]$-valued function on 2^V.
3. activation function of the form

$$f_v(W) = \begin{cases} 1 & \text{with probability } p_v(W) \\ 0 & \text{otherwise} \end{cases}$$

   where $W \subseteq S$ and every $w \in W$ is live at time t.
4. node v being activated in time t if $f_v(W) = 1$, where W is the set of neighbors of v live at time t.
5. the order-independence property, defined below.

   Note that each of the following definitions use p_v as an element of P and are defined over all $v \in V$. Likewise for f_v as an element of F defined over all $v \in V$.

Definition 2.9 The order-independence property states that when $\sigma : 1, \ldots, r \to 1, \ldots, r$ is a permutation function and $u_1, \ldots, u_r$ and $u_{\sigma_1}, \ldots, u_{\sigma_r}$ are two permutations of T, and $T_i = u_1, \ldots, u_{i-1}$ and $T_i' = u_{\sigma_1}, \ldots, u_{\sigma_{i-1}}$, then

$$\prod_{i=1}^{r}(1 - p_v(u_i \bigcup S \bigcup T_i)) = \prod_{i=1}^{r}(1 - p_v(u_{\sigma_i} \bigcup S \bigcup T_i'))$$

for all sets S disjoint from T.

   The probability of a node u influencing a node v depends on the set S of nodes that has already attempted to influence v. However, the ordering dependence property states that the probability of u activating v does not depend on the order of nodes in the set S that have previously attempted to activate v.

2.2.3.6 General Cascade and General Threshold Equivalence

The general cascade model has been shown to be equivalent to the general threshold model [94] under the following mapping:

1. for the probability function in the general cascade model:

$$p_u(u \bigcup S) = \frac{f_v(S \bigcup u) - f_v(S)}{1 - f_v(S)}$$

2. for the activation function in the general threshold model:

$$f_v(S) = 1 - \prod_{i=1}^{r}(1 - p_v(u_i) \bigcup S_{i-1})$$

where $S = u_1, \ldots, ur$ and $S_i = u_1, \ldots, u_i$.

This effectively says that by choosing the edge weights in either model, an instance of the general threshold model may be transformed into an instance of the general cascade model. This mapping ties the two models together and shows that diffusion is an equally hard problem on either model. Therefore, conclusions on one model also apply to the other model.

2.2.3.7 Decreasing Cascade Model

The decreasing cascade model was also defined by Kempe et al. [94] and is an extension of the general cascade model with the property that the more nodes that have attempted to activate a node, the less probability there is that the node becomes activated.

Definition 2.10 The decreasing cascade model is a diffusion model with all of the properties of the general cascade model with the additional property where $p_v(u \bigcup S) \geq p_v(u \bigcup T)$ whenever $S \subseteq T$.

2.2.3.8 Independent Cascade Model

This model was initially investigated by Goldenberg et al. in the context of marketing [73] and was defined by Kempe et al. [93]. Along with the linear threshold model, this model is classically used for studying diffusion on networks. It exists as a special case of the decreasing cascade model.

Definition 2.11 The independent cascade model is a diffusion model with all of the properties of the decreasing cascade model with the additional property that the $p_v(u \bigcup S) = p_v(u)$ for all sets $S \subseteq V$.

This means that the probability of a node u influencing a node v is independent of the set of nodes S that has attempted to influence v. Since we will be using this model for the remainder of our research, it is helpful to define some shorthand. We can look at this model as a set of edge probabilities on a graph.

Definition 2.12 On the independent cascade model, an edge probability, $b_{u,v}$ is the probability that a node u has to infect v whenever u is infected.

Note that $b_{u,v}$ does not necessarily equal $b_{v,u}$ and in fact, it will be the case in certain situations in our research that if $b_{u,v}$ is non-zero, that $b_{v,u}$ is 0.

It should be noted that the independent cascade model has the property that a node has exactly one time step in which it is infected to infect other nodes. That is, each node is infectious for exactly one time step and then can no longer be infected, nor can it infect any other nodes.

2.3   Problem Optimization

To better understand the underlying ideas behind diffusion and social networks, we study the formulations and optimizations for two important problems in social networks: (1) maximizing the spread of influence and (2) limiting the spread of misinformation, which is also called rumor blocking in some related work.

To begin with, we will cover some basic knowledge of social network. Social network is modeled as a directed graph $G = (V; E)$ with vertices in V modeling the individuals and edges in E modeling the relationship between individuals. For example, in co-authorship graphs, vertices are authors of academic papers and two vertices have an edge if the two corresponding authors have coauthored a paper.

2.3.1   Influence Maximization

An intensively studied problem in viral marketing is that, by picking a small group of influential individuals in a social network—say, convincing them to adopt a product—it will trigger the largest cascade of influence by which many users will try the product ultimately. Domingos and Richardson [53] are the first to pose it as a algorithmic problem and solve it as a probabilistic model of interaction. In [93], Kempe et al. formalize it as the problem of influence maximization.

A social network is modeled as a directed graph $G = (V, E)$ with vertices in V modeling the individuals and edges in E modeling the relationship between individuals. For example, in co-authorship graphs, vertices are authors of academic papers and two vertices have an edge if the two corresponding authors have coauthored a paper. Let p denote the influence probabilities between two vertices. The influence is propagated in the network according to a diffusion model m. Let S be the subset of vertices selected to initiate the influence propagation, which is also called seed set. Let $\sigma_m(S)$ be the expected number of influenced nodes at the end of propagation process. The formal definition of influence maximization problem is given as follows:

Problem 2.1 (Influence Maximization) Given a directed and edge-weighted social graph $G = (V, E, p)$, a propagation model m, and an integer $k \leq |V|$, find a seed set $S \subset V, |S| = k$, such that the expected influence $\sigma_m(S)$ is maximum.

This problem is also referred to as the identification of influential users or opinion leaders in a social network. This problem under both independent cascade (IC) and linear threshold (LT) propagation models is shown to be NP-hard [94], and so attempts have been made at approximating the value of $\sigma_m(S)$.

For a diffusion model with a non-negative, monotone submodular activation function, a greedy hill-climbing algorithm approximates the optimum within a factor of $(1 - 1/e) - \epsilon$ for any real number ϵ, as shown by Kempe et al. [93]. By greedy hill-climbing algorithm we mean an algorithm which, at every step, adds to the output set the node that currently has the highest influence spread. The challenge of the greedy algorithm rises when selecting a new vertex v that provides the largest marginal gain $\sigma_m(S + v) - \sigma_m(S)$ compared to the influence spread of current seed set S. Computing the expected spread given a seed set turns out to be a difficult task under both the IC model and the LT model. Instead of finding an exact algorithm, Kempe et al. run Monte Carlo simulations of the propagation model for sufficiently many times (10,000 trials) to obtain an accurate estimate of the influence spread, leading to a very long computation time.

A vast number of papers have studied improving the efficiency and availability of the influence maximization [25, 37, 39, 130, 149, 183, 187, 188]. In [37], Chen et al. also propose a degree discount heuristics with influence spreads and combines a Cost-Effective Lazy Forward (CELF) scheme to further improve the greedy algorithm. In [39], Chen et al. propose a scalable heuristic called DAGs (local directed acyclic graphs) for the linear threshold model. They construct local DAGs for each node and computing the expected spread over DAGs can be done in linear time while over general graphs it is #P-hard. In [130], Mathioudakis et al. simplified the network to accelerate the speed of finding seeds. However, these heuristics lack of theoretical guarantees. At this front, the state of the art is the reverse influence sampling (RIS) approach [25, 188]. These methods attempt to generate a $a1 - 1/ - \epsilon$ approximation solution with minimal number of RIS samples. And the IMM algorithm [188] is among the most competitive ones so far. In [149], Nguyen et al. generalize the RIS sampling methods into sampling frameworks and optimize it by an innovative stop and share strategy. Their method uses minimum number of samples while meeting strict theoretical thresholds for the influence maximization problem.

Another issue for Kempe's method is that it assumes a weighted social graph as input and does not address the problem of learning influence probabilities. In [164], Saito et al. study how to learn the probabilities of the IC model from a set of past propagations by formalizing this as a likelihood maximization problem and then applying the expectation maximization (EM) algorithm to solve it; Goyal et al. [75,

76] propose a credit model for learning influence probability from pure historical action logs which takes the temporal nature of influence into account. In [207], Xu et al. first present a method to identify influential entities in large social networks based on a weighted maximum cut framework which is totally separate from traditional method of greedy strategy while maintaining high efficiency. Moreover, they have developed a new method of learning influence strength by analyzing both social network structure and historical user data.

Some variations are proposed to handle different real-world requirements, such as looking at communities, competitive and complementary influence maximization. Leskovec et al. [115, 183] optimized placements for a set of social sensors such that the propagation of information or virus can be effectively detected in a social network. Lappas et al. [111] discover a set of key mediators which determine the bottlenecks of influence propagation if seed nodes try to activate some target nodes. Sun et al. [183] study the multi-round influence maximization problem, where influence propagates in multiple rounds independently from possibly different seed sets.

A characteristic common to the studies discussed so far is the assumption that information cascades of campaigns happen in isolation. Next we introduce a group of problem formulations that capture the notion of competing campaigns in a social network [19, 26, 33, 40, 104, 190]. This scenario will frequently arise in the real world: multiple companies with comparable products will vie for sales with competing word-of-mouth cascades; similarly, many innovations face active opposition also spreading by word of mouth. Carnes et al. [33] study the strategies of a company that wishes to invade an existing market and persuade people to buy their product. This turns the problem into a Stackelberg game where in the first player (leader) chooses a strategy in the first stage, which takes into account the likely reaction of the second players (followers). In the second stage, the followers choose their own strategies having observed the Stackelberg leader decision, i.e., they react to the leader's strategy. Carnes et al. use models similar to the ones proposed in [93] and show that the second player faces an NP-hard problem if aiming at selecting an optimal strategy. Furthermore, the authors prove that a greedy hill-climbing algorithms leads to a $(1 - 1/e - \epsilon)$-approximation.

Around the same time, Bharathi et al. [19] introduce roughly the same model for competing rumors and they also show that there exists an efficient approximation algorithm for the second player. Moreover they present an FPTAS for the single player problem on trees. Kostka et al. [104] considered the rumors diffusion as a game theoretical problem under a much more restricted model compared with IC and LT. They showed that the first player did not always obtain benefit although he/she started earlier. Trpevski et al. [190] propose a competitive rumors spreading model based on SIS model in epidemic domain, but they did not address the issue of influence maximization or rumor blocking. Borodin et al. in [26] study competitive influence diffusion in several different models extended from LT. Chen et al. [40] address positive influence maximization under an extension of the IC model with negative opinions about the product or service quality.

2.3.2   Misinformation Minimization

While the ease of information propagation in social networks can be very beneficial, it can also have disruptive effects. A number of examples of this sort are the spread of misinformation on swine flu in Twitter [135], exaggerated reports on a bomb attack in Grand Central, and celebrities that are falsely claimed as being dead [86]. We specifically focus on the study that addresses the problem of influence limitation [30] where a bad campaign starts propagating from a certain node in the network and use the notion of limiting campaigns to counteract the effect of misinformation. The problem of misinformation minimization can also be called as rumor blocking problem or influence limitation problem. Its definition is defined as follows.

Problem 2.2 (Misinformation Minimization) Given a graph $G = (V; E; p)$, where p represents its positive and negative edge weights, a negative seed set N_0, and a positive integer k, the goal is to find a positive seed set S of size at most k such that the expected number of negatively activated nodes is minimized, or equivalently, the reduction in the number of negatively activated nodes is maximized.

Kimura et al. in [98] deal with influence limitation problem through blocking a certain number of links in a network. The most recent works regarded with this problem include [30, 84, 147]. In [30], Budak et al. study the controlling of negative information in social networks, that is, when negative information is diffused in networks, how to select some nodes to implant positive information in order to correct the information attitude in the whole network to a maximizing extent. They prove that under an extension of the IC model, the eventual influence limitation (EIL) problem is NP-hard. They also examine a more realistic problem of influence limitation in the presence of missing information and introduced an algorithm called predictive hill-climbing approach which has good performance.

In [84], He et al. propose a competitive linear threshold (CLT) model to address the influence blocking maximization (IBM) problem, which is an extension to the classic linear threshold model. They prove that this problem under CLT model was submodular and theoretically obtained a $(1 - 1/e)$-approximation ratio by a greedy strategy. To improve the efficiency, they further propose the CLDAG algorithm that is similar to the LDAG algorithm in [39]. In [147], a β_T^I-Node Protector problem is proposed by Nguyen et al., which is actually the extensions of the misinformation minimization problem under LT and IC models. The goal is to find the smallest set of highly influential nodes that can limit the viral spread of misinformation originated from set I to a desired rate $(1 - \beta)$ $(\beta \in [0, 1])$ in T time steps. They present a greedy viral stopper (GVS) algorithm that greedily adds nodes with the best influence gain for β Node Protectors to the current solution. They also apply GVS to the network restricted to T-hop neighbors of the initial set I and reached a slightly better bound for β_T^I-Node Protector problems. Besides, a community based algorithm which returns a good selection of nodes to decontaminate in a timely manner is proposed.

2.4   Large-Scale Data Analysis

No matter which technique is used in studying information diffusion, large-scale data analysis is a significant aspect of study as well as being a significant challenge. In this part, we will introduce several representative data analysis techniques used in the social influence analysis. With the increase of studies in social networks, there are a number of datasets available to researchers [109, 113, 146].

As data grows, data mining and machine learning applications start to embrace the Map-Reduce paradigm, e.g., news personalization with Map-Reduce EM algorithm [49], Map-Reduce of several machine learning algorithms on multicore architecture [45]. For the networking data, graphical probabilistic models are often employed to describe the dependencies between observation data. Markov random field [180], factor graph [105], restricted Boltzmann machine (RBM) [201], and many others are widely used graphical models. In [186], Tang et al. proposed a topical factor graph (TFG) model, for quantitatively analyzing the topic based social influences. Compared with the existing work, the TFG can incorporate the correlation between topics. They also proposed a very efficient algorithm for learning the TFG model. In particular, a distributed learning algorithm has been implemented under the Map-Reduce programming model.

The techniques used in Web community discovery can also be applied in social influence analysis. The problem of detecting such communities within networks has been well studied. Early approaches such as spectral partitioning, the Kernighan-Lin algorithm, hierarchical clustering, and G-N algorithm work well for specific types of problems (particularly graph bisection), but perform poorly in real networks. Recently, most works focus on graph partitioning approaches. The most popular partition technique in the literature is k-means clustering, which aims to separate the nodes in k clusters such to maximize/minimize a given cost function based on distances between nodes and/or from nodes to centroids. In [209], Q. Yan et al. proposed a two-phase method that combines community detection with naive greedy algorithm to improve time efficiency of influence maximizing problem with multiple spread model. In the first phase, they use efficient clustering algorithm such as kernel k-means to partition graph nodes into k clusters, with the parameter k related to the number of influential nodes. In the second phase, in each community, they apply techniques in social influence maximization to find influential nodes in each cluster. Similar work has [48].

2.5   Research Trends

Social networks provide large-scale information infrastructures for people to discuss and exchange ideas about different topics. The general problem of network influence analysis represents a new and interesting research direction in social network mining. There are many potential future directions of this work. Even though the influence diffusion in social networks has been intensively studied, we note that there are three essential dimensions emerging from the analysis we performed, which could be of great benefits for future researchers.

2.5.1   Learn Influence Probabilities of Diffusion Models

In social network analysis, two information diffusion models: the independent cascade (IC) and the linear threshold (LT) are widely used to solve such problems as the influence maximization problem and the misinformation minimization problem. These two models focus on different information diffusion aspects. The IC model is sender-centered (push type) and each active node independently influences its inactive neighbors with given diffusion probabilities. The LT model is receiver-centered (pull type) and a node is influenced by its active neighbors if their total weight exceeds the threshold for the node. What is important to note is that both models have parameters that need be specified in advance: diffusion probabilities for the IC model, and weights for the LT model. However, their true values are not known in practice. This poses yet another problem of estimating them from a set of information diffusion results that are observed as time sequences of influenced (activated) nodes. This falls in a well-defined parameter estimation problem in machine learning framework.

In [165], K. Saito et al. extended both IC and LT models to be able to simulate asynchronous time delay. They learned the dependency of the diffusion probability and the time delay parameter on the node attributes by solving a formulated problem named as the maximum likelihood estimation problem, and an efficient parameter update algorithm that guarantees the convergence is derived. Other efforts of learning parameters of the influence graph from history data include the work [75, 162]. In [75], A. Goyal et al. proposed both static and time-dependent models for capturing influence. Moreover, they presented optimized algorithms for learning the parameters of the various models based on social networks and historical action logs.

2.5.2   Learn the Speed of Influence Spread in Networks

It has been observed that information spreads extremely fast in social networks. There has been some but not enough theoretical results about the analysis of influence spread speed. In [52], B. Doerr et al. have shown that for preferential attachment graphs the classic push-pull strategy needs $\Theta(logn)$ rounds to inform all vertices. The slightly improved version which avoids that a vertex contacts the same neighbor twice in a row only needs $\Theta(logn/loglogn)$ rounds, which is best possible since the diameter is of the same order of magnitude. In [66], N. Fountoulakis et al. establishes for a class of random graphs ultrafast time bounds on the running time of the synchronous push-pull protocol that is needed until the majority of the vertices are informed. They present the first theoretical analysis of this protocol on random graphs that have a power-law degree distribution with an arbitrary exponent $\beta > 2$. Their main findings reveal a striking dichotomy in the performance of the protocol that depends on the exponent of the power law. More specifically, it is shown that if $2 < \beta < 3$, then the rumor spreads to almost all nodes in $\Theta(loglogn)$ rounds with high probability. On the other hand, if $\beta > 3$, then $\Theta(logn)$ rounds are necessary.

2.5.3   Study Variations of Influence Maximization

Traditional diffusion models including IC and LT do not fully incorporate important temporal aspects that have been well observed in the dynamics of influence propagation. Firstly, the propagation of influence from one person to another may incur a certain amount of time delay, which is obvious from recent studies by statistical physicists on empirical social networks. Secondly, the spread of influence may be time-critical in practice. In a certain viral marketing campaign, a company might wish to trigger a large cascade of product adoption in a fairly short time frame, e.g., a 3-day sale. Therefore it is very meaningful to extend the influence maximization problem to have a time constraint.

Chen et al. [41] proposed the time-critical influence maximization problem, in which one wants to maximize influence spread within a given deadline. In their model influence delays are constrained to follow the geometric distribution. In [119], B. Liu et al. proposed a new problem of the time constrained influence maximization in social networks based on a latency aware independent cascade model. They also proposed to use influence spreading paths to quickly and effectively approximate the time constrained influence spread for a given seed set. Sun et al. [183] propose to study multi-round influence maximization problem, which models the viral marketing scenarios in which advertisers conduct multiple rounds of viral marketing to promote one product.

Chapter 3
Information Source Detection in Social Networks

3.1 Introduction

The rising popularity of online social networks has made information generating and sharing much easier than ever before, due to the ability to publish content to large, targeted audiences. Such networks enable their participants to simultaneously become both consumers and producers of content, shifting the role of information broker from a few dedicated entities to a diverse and distributed group of individuals. While this fundamental change allows information propagating at an unprecedented rate [166], it also enables unreliable or unverified information spreading among people, such as rumors [47].

Rumor has been a research subject in psychology and social cognition for a long time [50]. It is often viewed as an unverified account or explanation of events circulating from person to person and pertaining to an object, event, or issue in public concern [152]. Bordia et al. [23] propose that transmission of rumor is probably reflective of a "collective explanation process." Since there is often not enough resource to manually identify rumors or misinformation from the huge volume of fast evolving data, it has become a critical problem to design systems that can automatically detect misinformation and disinformation. Microblogging services, like Twitter, allow small pieces of information to spread quickly to large audiences, allowing rumors to be created and spread in new ways [108].

Current media environment is suitable to the emergence and propagation of rumors that are not limited to insignificant subjects: Rumors can have major consequences on political, strategic, or economical decisions. Increasingly, they are triggered off on purpose for various reasons: campaigns can be carried out in order to discredit a company, endanger strategic choices, or question political decisions. Therefore research on rumor detection has great significance on Web security issues [140].

In recent times, many Web based systems have been developed to detect and evaluate the rumors in social networks. Examples are (1) TwitterTrails.com [133], a

© The Author(s), under exclusive license to Springer Nature Switzerland AG 2020
W. Xu, W. Wu, *Optimal Social Influence*, SpringerBriefs in Optimization,
https://doi.org/10.1007/978-3-030-37775-5_3

system which permits users to determine the features of propagated rumors and its falsification, (2) TweedCred [80], an instantaneous system to judge trustworthiness of posts on Twitter, (3) Hoaxy [174], a platform for tracking the misinformation in a social network, (4) Emergent.info [189], a real-time rumor follower that focused on rising tales on the Internet and observes their faithfulness, and (5) Snopes.com [181] and factcheck.org [60], admired websites archiving memes and urban myths. The reality checking abilities of these rumor detection systems validate the authentication of rumors on Web and vary from entirely automatic to semi-automatic. But, these systems do not track or observe the diffusion progress and do not detect all possible source(s).

Rumor Detection using Machine Learning Social network analysis about studying rumors often focuses on machine learning techniques such as building classifiers, sentiment analysis, Twitter data mining, and so on. Work in this area includes [59, 116, 158, 160]. Leskovec et al. use the evolution of quotes reproduced online to identify memes and track their spread overtime [116]. Ratkiewicz et al. [160] created the Truthy system, identifying misleading political memes on Twitter using tweet features, including hashtags, links, and mentions. Other projects focus on highlighting disputed claims on the Internet using pattern matching techniques [59]. Qazvinian et al. [158] explore the effectiveness of three categories of features: content based, network based, and microblog specific memes for correctly identifying rumors in microblogs. In these introduced research work, a complete set of social conversations (e.g., tweets) that are actually about the rumor need to be retrieved first.

There have appeared some studies on analyzing rumors and information credibility on Twitter, the world's largest microblogging platform. Castillo et al. [34] focus on automatically assessing the credibility of a given set of tweets. They analyze the collected microblogs that are related to "trending topics," and use a supervised learning method (decision tree) to classify them as credible or not credible. Qazvinian et al. [158] focus on two tasks: The first task is classifying those rumor-related tweets that match the regular expression of the keyword query used to collect tweets on Twitter monitor. The second task is analyzing the users' believing behavior about those rumor-related tweets. They build different Bayesian classifiers on various subsets of features and then learn a linear function of these classifiers for retrieval of those two sets. Mendoza et al. [132] use tweets to analyze the behavior of Twitter users under bombshell events such as the Chile earthquake in 2010. They analyze users' retweeting topology network and find the difference in the rumor diffusion pattern on Twitter environment than on traditional news platforms.

Rumor Source Detection Based on Information Spreading Many studies on the problem of information propagation are inspired from the more common issue of contagion and generally use models for viral epidemics in populations such as the susceptible-infected-recovered (SIR) model. On this subject, research has focused on the effects of the topological properties of the network on inferring the source of a rumor in a network. Shah and Zaman [171–173] were the first to study systematically the problem of infection sources estimation which consider an susceptible-infected (SI) model, in which there is a single infection source, and susceptible nodes are those with at least one infected neighbor, while infected nodes do not recover. Subsequently, [123, 126] consider the multiple sources estimation problem under an SI model; [211] studies the single source estimation problem for the susceptible-infected-recovered (SIR) model, where an infected node may recover but can never be infected again; and [125] considers the single source estimation problem for the susceptible-infected-susceptible (SIS) model, where a recovered node is susceptible to be infected again.

Although all the works listed above answer some fundamental questions about information source detection in large-scale networks, they assume that a complete snapshot of the network is given while in reality a complete snapshot, which may have hundreds of millions of nodes, is expensive to obtain. Furthermore, these works assume homogeneous infection across links and homogeneous recovery across nodes, but in reality, most networks are heterogeneous. For example, people close to each other are more likely to share rumors and epidemics are more infectious in the regions with poor medical care systems. Therefore, it is important to take sparse observations and network heterogeneity into account when locating information sources. In [124, 170, 212], detecting information sources with partial observations in which only a fraction of nodes (observers) can be observed has been investigated. The work in [154] assumes that for each of the observers, the knowledge of the first infected time and from which neighbor the infection is received are given. This assumption is impractical in some cases. For example, it is usually hard to know from which neighbor the infection is coming from in a contagious disease spreading within a community. In [54], the authors have considered the detection rate of the rumor centrality estimator when a priori distribution of the source node is given. Several other source locating algorithms have also been proposed recently, including an eigenvalue based estimator [157], a fast Monte Carlo algorithm [2], and a dynamic message-passing algorithm under the SIR model [2].

Source detection is very significant in various application domains such as medical (to find the source of epidemic), security (to detect the source of virus), large interconnected network (to detect the flaws in power grid network, gas or water pipeline network), social network (to identify the culprits who spread wrong information), financial network (for checking the reasons of cascade failures), etc. Due to its wide scope in different applications, past two decades observed large improvements in source detection techniques. Major research has been done for source identification in different application areas like finding the first patient to control an epidemic of disease [9], source of virus [171], gas leakage source in wireless sensor network [177], propagation sources in complex networks [90], and source of rumor or misinformation in a social network [148, 175, 206, 210] which are directly or indirectly related to rumor source detection.

In Sect. 3.2, we introduce a monitor based approach to detect single rumor source in online social networks and define a probability-score function, named as rumor quantifier, for ranking how likely nodes are the actual rumor source [206]. Given a weighted social graph, we propose a polynomial time algorithm to detect the rumor source with respect to the popular independent cascade (IC) model in online social networks. Since real social networks are getting bigger with millions of nodes, our algorithm can scale well on real large datasets.

In Sect. 3.3, we consider detecting multiple rumors from a deterministic point of view by modeling it as the set resolving set (SRS) problem [210]. Let G be an undirected graph on n vertices. A vertex subset K of G is SRS of G if any different detectable node sets are distinguishable by K. The problem of multiple rumor source detection (MRSD) will be defined as finding an SRS K with the smallest cardinality in G. Using an analysis framework of submodular functions, we propose a highly efficient greedy algorithm for MRSD problem on general graphs, which is polynomial time under some reasonable constraints, that is, there is a constant upper bound r for the number of sources. Moreover, we show that our natural greedy algorithm correctly computes an SRS with provable approximation ratio of at most $(1 + r \ln n + \ln \log_2 \gamma)$, given that γ is the maximum number of equivalence classes divided by one node-pair. This is the first work providing explicit approximation ratio for the algorithm solving minimum SRS. Therefore the introduced framework suggests a robust approach for MRSD independent of diffusion models in networks.

In the last section, we will summarize our work of rumor source detection and future work in this field will also be discussed.

3.2 Single Source Detection

This section studies the problem of identifying source location of single rumor in online social networks in which the spread of information follows the popular independent cascade (IC) model. In the absence of text information, we develop a monitor based approach to evaluate how likely that a piece of information is actually a rumor. Given the underlying social network structure, a number of monitor nodes are injected into the network whose job is to report the data they receive. Based on observing which of monitors received the information and which did not, we propose a polynomial time algorithm to compute rumor quantifier, a reachability based score for ranking the importance of nodes as the rumor source. Extensive simulation results have shown that, with a reasonable number of monitor nodes and appropriate monitor deployment, our rumor source detection algorithm can recognize rumor source effectively and efficiently.

3.2.1 Problem Formulation

We first introduce the propagation model of rumors, then present the formal problem formulation. Opinion dynamics in a social network can be modeled in some cases using independent cascade (IC) Model, which is a classical model in influence spreading. A individual on Twitter may be influenced by the opinion or posting of someone she is following, thereby becoming "infected" with the same opinion.

Independent Cascade Model A social network is modeled as a directed graph $G = (V, E)$ where V is the set of users and E is the set of edges where each edge represents relationship between two individuals. Let $u \in V$ be a rumor source form which a rumor starts spreading at time, say t. As basic independent cascade (IC) model operates, the cascade of rumor spreads in the graph. Specifically, u is given a single chance to activate each currently inactive neighbor v; it succeeds with a probability $p(u, v)$—a parameter of the system—independently of the history. (If u has multiple inactive neighbors, its attempts are sequenced in an arbitrary order.) If u succeeds, then v will become active in step $t+1$. Again, the process runs until no more activations are possible. In this model, once a node is infected with the rumor, it retains it forever.

Note that cascades in IC model are necessarily trees since if a node, say s, gets infected multiple times knowing the node that infected s first is sufficient. Thus, the influence structure of a cascade is given by a directed tree T, which is contained in the directed graph G, i.e., the graph over which the cascade propagates.

Problem Definition Given the above spreading model, the goal of rumor source detection is to identify the rumor source based on the input weighted social graph.

Monitor We assume that a set of k pre-selected nodes M ($M \in V$) are our monitors. For rumor investigation purposes, given a specific piece of information (cascade), monitors report whether they have received it or not. We denote the set of monitor nodes who have received the rumor by M^+, and the set of monitor nodes who have not received it by M^- (where $M^+, M^- \in M$). We call the former set positive monitors and the latter negative monitors.

Social Influence Probability Social influence from node u to v, denoted by $p(u, v)$, is a numerical weight associated with the edge $e \in E$. In most cases, the social influence score is asymmetric, i.e., $p(u, v) \neq p(u, v)$. Furthermore, the social influence from node u to v will vary on different types of networks.

Thus based on the above concepts, we can define the tasks of rumor source detection. In this paper, we only discuss the case that there is only one rumor source. Given a weighted social network $G = (V; E; p)$, a propagation model, and a monitor set, the goal is to identify the source node that starts the rumor cascade. The problem definition is as follows.

Problem 3.1 (Single Rumor Source Detection) Given an cascade model m, we observe the rumor infected graph $G = (V, E, p)$ at some time $t > 0$. We do not know the value of t or the realization of the spreading times on edges $e \in E$; we only know positive monitors $M^+ \subseteq V$ and negative monitors $M^- \subseteq V$. The goal is to find the rumor source $r \in V$ given G.

3.2.2 Monitor Based Approach

Our main idea is to leverage monitors for rumor source detection. Firstly, we introduce a rumor quantifier $Q(r)$, a probability-score function, for ranking the importance of nodes as the rumor source. Let $Q(r)$ denote how likely that a node $r \in V$ is actually the rumor source. Identifying the rumor source will be formulated as finding a node $r \in V$ that maximizes the rumor quantifier $Q(r)$, which is written as follows:

$$Max \quad Q(r) \tag{3.1}$$

To identify the source of a rumor, we use the intuition that information will spread more quickly from the source to the positive monitors but more slowly to the negative monitors. Since our model is probabilistic and dynamic, in other words, the cascade must be easier to propagate from the source to positive monitors while harder to propagate from the source to negative monitors. Based on this idea, the equation of $Q(\cdot)$ can be defined as follows:

$$Q(r) = F(p(r, M^+), p(r, M^-)) \tag{3.2}$$

where $p(r, M^+)$ denotes the probability that a cascade spreads from node r to the positive monitor set M^+, and $p(r, M^-)$ denotes the probability that the cascade spreads from node r to the negative monitor set M^-.

The function F demonstrates that the quantifier $Q(r)$ of a node r is affected by two factors: both $p(r, M^+)$ and $p(r, M^-)$. For each node $r \in V$, the quantifier first considers how likely that a cascade spreads from it to the positive monitor set, $p(r, M^+)$. In specific, the larger of the probability of $p(r, M^+)$, the larger of the value of $Q(r)$. In other word, the value of our quantifier is proportional to the value of $p(r, M^+)$. Also, the smaller of the value of $p(r, M^-)$, the larger of the value of $Q(\cdot)$. Thus, the node with maximum rumor quantifier $Q(\cdot)$ has the maximum likelihood estimation in the context of independent cascade model. The details of the function F will be further depicted in the section of algorithm.

For a cascade, we will first specify the influence probability $p(u, v)$ that describes how likely that node u spreads the cascade to node v. Then we will describe the probability $p(r, M^+)$ which specifies the probability that the cascade propagates from node r to the positive monitor set M^+. Similarly, we also define $p(r, M^-)$, which describes how likely cascade propagates from node r to the negative monitor set M^- in the network G.

For a path $P = < p_1, p_2, \ldots p_i, \ldots, p_m >$, we define the propagation probability of the path P as,

$$p(P) = \prod_{i=1}^{m-1} (p_i, p_{i+1}) \tag{3.3}$$

where the product is over the edges of path P. Intuitively the probability that u activates v through path P is $p(P)$, because it needs to activate all nodes along the path. Here, the edges of the path P simply specify how the cascade spreads, i.e., every node gets influenced by its parent.

To approximate the actual expected influence within the social network, we propose to use the maximum influence path (MIP) to estimate the influence from one node to another. Let $P_G(u, v)$ denote the set of all paths from u to v in a graph G.

Definition 3.1 (Maximum Influence Path (MIP))   For a social graph G, we define the Maximum Influence Path $MIP(u, v)$ from u to v in G as

$$MIP(u, v) = argmax_P\{p(P)|P \in P_G(u, v)\}.$$

Ties are broken in a predetermined and consistent way, such that $MIP(u, v)$ is always unique, and any subpath in $MIP(u, v)$ from x to y is also the $MIP(x, y)$. If $P_G(u, v) = \varnothing$, we denote $MIP(u, v) = \varnothing$.

For any two nodes $u, v \in V$, if there exists no path connecting from u to v, then $p(u, v) = 0$ since they cannot influence each other. Otherwise suppose there exists multiple paths connecting from u to v, we define $p(u, v)$ according to maximum influence path as follows.

$$p(u, v) = MIP(u, v) \qquad (3.4)$$

Now that we have specified the probability $p(u, v)$ for any two nodes $u, v \in V$, next we define the probability of observing cascade propagating from a node r to a monitor set M in a particular tree structure T as

$$p_T(r, M) = \phi_{m_i \in M^+} p(r, m_i) \qquad (3.5)$$

A typical way of function ϕ is to summarize $p(r, m_i)$ for all the nodes $m_i \in M$. Suppose $|M| = n$, we provide a heuristic in our experiment for function ϕ, as demonstrated in Eq. (3.6) such that $p(u, M)$ will not exceed 1.

$$\phi_{m_i \in M} p(r, m_i) = 1 - \prod_{i=1}^{i=n}(1 - p(r, m_i)) \qquad (3.6)$$

Here we will use a specific example to illustrate Eq. (3.6). Figure 3.1 shows the propagation tree form root node u to the monitor set, say M^+, which has three nodes: m_1, m_2, m_3. The influence probabilities among nodes are given. According to Eq. (3.6), $P(u, M+) = 1 - (1 - 0.8 * 0.8 * 0.3)(1 - 0.8 * 0.6)(1 - 0.5) = 0.790$.

In this section, we have introduced the rumor quantifier $Q(\cdot)$ for ranking the probability that a node is the rumor source and how to compute it. In the following section, we will develop efficient algorithms to detect the rumor source.

Fig. 3.1 An example of propagation tree from root node u to monitor node set $\{m_1, m_2, m_3\}$. The influence probabilities are given as edge weights

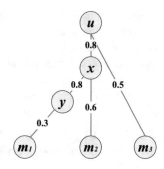

Algorithm 1 Maximum propagation tree identification

Input: Root Node r, Leaf Nodes $m_i \in M$, Weighted Social Graph $G = (V, E, p)$.
Output: Maximum Propagation Tree T

 Set $p' = -lnp$.
 $G' = (V, E, P')$.
 for all node $m_i \in M$ **do**
  Find the shortest path h from r to m_i in G'.
 end for
 return all the paths h.

3.2.3 Proposed Algorithm

Given a node r and a monitor node set M, there are more than one propagation trees satisfying the condition that the root is r and the leaf node set is M. To reduce the computation complexity, instead of searching all the possible propagation trees, we only consider the most likely propagation tree from r to M. Here we give a formal definition of the most likely propagation tree, named as maximum propagation tree.

Definition 3.2 (Maximum Propagation Tree (MPT)) Given a weighted social graph $G = (V, E, p)$, a root node r, a monitor node set M, a maximum propagation tree ($MPT(r, M)$) consists of all the maximum influence paths from root node r to each node in the set M.

 In order to find the maximum propagation tree given a root node, leaf nodes (the monitor set) and the underlying graph structure, we propose a polynomial time solution, as demonstrated in Algorithm 1. The general idea is to find all the shortest paths from the root node to all the leaf nodes.

 Next we will give a proof that why Algorithm 1 can find the desired maximum propagation tree.

Proof Let $G = (V, E, p)$ be a weighted social graph. In terms of problem definition of MPT, the maximum propagation tree for G is a tree, say $T \subseteq G$, which consists of all the maximum influence paths from root node r to each node in the set M. In Algorithm 1, all the weights p is updated to $p' = -lnp$. For each node $m_i \in$

M^+, if a shortest path from r to m_i is found, then $p'(r, m_i) = -lnp$ is minimized. Since $p(r, m_i) \in (0, 1)$ (influence probability), $lnp(r, m_i) < 0$, $p' = -lnp > 0$, thus, $p(r, m_i) = \prod_{(i,j) \in edge connecting r, m_i} (i, j)$ is maximized. In other word, the maximum influence path from root node r to $m_i \in M^+$ is actually the shortest path between them.

To detect the hidden rumor source, our algorithm is to, for every mode $u \in V$, first calculate the rumor quantifier $Q(u)$, that is $p(u, M^+)$ in our settings here. Rank nodes according to the value of rumor quantifier decreasingly. The node with maximum $Q(u)$ is most likely to be the rumor source. The detailed algorithm is described in Algorithm 2. Given a weighted social graph G in which the weight denote the influence probability between individuals, and monitor sets M^+, M^-, for each node u, we want to find a maximum propagation tree $T \in G$ such that the root is set as node u, while the leafs of the tree are predefined as positive monitors (line 2). Based on the found maximum propagation tree, the algorithm computes probabilities $p(u, M^+)$ (line 3). The reason why such tree exists is that, if a node u is the rumor source, u must have paths to all the monitors in M^+. Otherwise u cannot be a rumor source. When it comes to a special case that several nodes cannot be distinguished by positive monitor set (say, nodes have the same probability of $p(v, M^+)$) (line 5), the quantifier will consider how likely the cascade spread from them to negative monitor set (line 6). In this case, the node with lower probability of $p(v, M^-)$ will be ranked higher. Therefore, our rumor quantifier relies mainly on the positive monitor set while at the same time it does not neglect the effect of negative monitor set.

3.2.4 Experiments

To test our monitor based method to identify the rumor source, we run our RSD (**R**umor **S**ource **D**etection) algorithm on graphs of a real online social network. We are interested in understanding its behavior in practice and comparing its

---

Algorithm 2 Rumor source detection

Input: Monitor set M^+, M^-, weighted social Graph $G = (V, E, p)$.
Output: Rumor Source node.
   for all node $u \in V$ **do**
      $T =$ Maximum Propagation Tree Identification (u, M^+, G);
      Compute the probability $p(u, M^+)$ based on T;
   end for
   for nodes $v \in V$ with the same probability of $p(v, M^+)$ **do**
      Rank them by $p(v, M^-)$ increasingly;
   end for
   return the node ranking first.

---

performance under various monitor deployment. We find that our RSD algorithm achieves significant accuracy (up to 87%) over real datasets of social networks.

3.2.4.1   Dataset

At September 2010, Twitter reports that its users publish nearly 95 million tweets per day [194]. This makes Twitter an excellent case to analyze rumors in social media. In our experiment, we extracted a social graph structure from Twitter using Twitter search API as testbed. The data we collect has 38,484 nodes, 1,364,322 edges where the nodes represent users, the edges represent the friendship or followership among the users. Besides the topology, we also calculated Retweet probability of each edge $x \rightarrow y$ as the ratio of x's tweets retweeted by y to all tweets of x. Calculated retweet probabilities were used to simulate independent cascade propagation of rumors.

3.2.4.2   Experimental Evaluation

The goal of the experiments on synthetic data is to understand how the underlying network structure and monitor deployment affect the performance of our algorithm. In general, we proceed experiments as follows: (1) given a weighted social graph extracted from Twitter, we simulate a cascade (a random rumor source is selected); (2) using the retweet probability of each edge, the rumor is propagated according to IC diffusion model; (3) if the rumor fails to reach 1% of all nodes, it is viewed as a negligible rumor and this rumor propagation is discarded. A new rumor is selected and the same process is repeated. For each rumor propagation (cascade), we try to identify the rumor source using our RSD algorithm with different number of monitors. To compute average precision of the algorithm, we simulate cascades 200 times.

For best accuracy, it is important to choose monitors wisely. In this paper, we compare the following three monitor selection methods.

Random   Random selection method selects k monitors randomly. This means that, for any node $x \in V$, the probability that x is selected as a monitor is $k/ \mid V \mid$.

Incoming Degree (ID)   In this method, the number of incoming edges of each node is counted. Then, the top k nodes which have largest counts are chosen as monitors.

Betweenness Centrality (BC)   This method calculates betweenness centrality [145] for each node v, which is defined as

$$C(v) \sum_{s \neq t \neq v \in V} \frac{\sigma_{st}(v)}{\sigma_{st}}$$

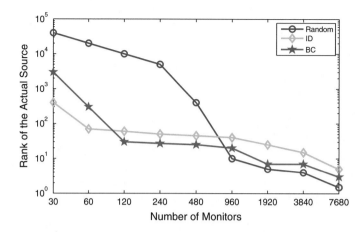

Fig. 3.2 Average rank of the actual rumor source in the output (out of 38,484 nodes)

where σ_{st} is the number of shortest paths from s to t and pass through v. Then, the k nodes which have the largest betweenness centrality are chosen as monitors.

3.2.4.3 Experimental Results

Using the method presented in Sect. 3.3, all nodes are sorted in the likelihood that they are the actual rumor source. Figure 3.2 shows the average rank of the actual source in the output. In the ideal case, the rank should be one which means that the top suspect is actually the rumor source. Note that, regardless of the monitor selection method, the rank of the true source generally decreases (i.e., improves by becoming closer to 1) as the number of monitors increases. Here the monitor numbers are set increasingly as [30, 60, 120, 240, 480, 960, 1920, 3840, 7680]. At first when the number of monitor nodes is small, Random performs worst, but it improves as more monitors are added. In contrast, Incoming Degree (ID) performs quite well when there are small number of nodes but is not satisfactory when the number of monitors is very large. The performance of betweenness centrality (BC) lies in between.

Figure 3.3 shows the distance between the top suspect and the actual source of all monitor selection methods. Note that no matter how many monitors are chosen, the average distance is smaller than three steps. In the ideal case, the distance should be zero, meaning the top suspect is the source. Figure 3.3 shows a similar tendency as Fig. 3.2. The distance decreases as more monitors are added. Random has large distances with a small number of monitors, but the distance decreases drastically as the number of monitors increase. BC and ID generally show the smallest distance between the top suspect and the actual source.

We also observe the details of monitors except the number and deployment. Figure 3.4 shows the ratio of experiments in which no monitor received the rumor.

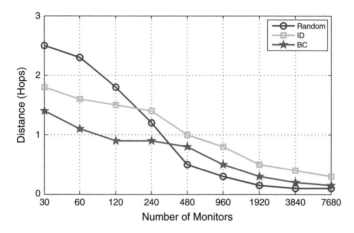

Fig. 3.3 Average distance between the found rumor source and the actual rumor source

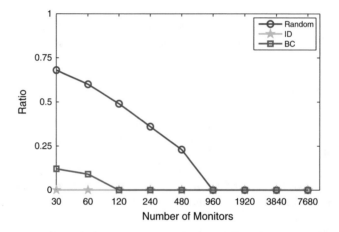

Fig. 3.4 Ratio of experiments in which no monitor receives a rumor (out of 200 cascades)

In all monitor selection methods, the ratio decreases as the number of monitors increases. Among the three methods compared, the Random selection method has the highest ratio. When the number is monitors is small, Random tends to choose nodes loosely scattered on the boundary of the graph. Therefore, monitors selected by Random have low probability of hearing rumors. The Random selection method also has a high ratio of negative monitors when the number of monitors is small. The other methods (ID, BC) have small ratio compared to Random. When no monitor hears the rumor, it is very hard to find the source accurately as shown in Fig. 3.2 (Random when the number of monitors is 30, for example).

Now we take a detailed look about the accuracy of our RSD algorithm. As Fig. 3.5 demonstrates, the precision of our algorithm increases with the monitor number increases. The precision is defined as the number of experiments when the

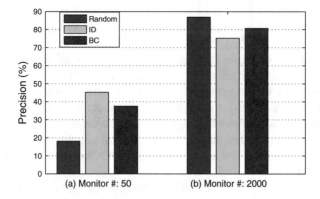

Fig. 3.5 Average precision of RSD algorithm in finding the actual rumor source (out of four hundreds of experiments). The monitor numbers are set as 50 and 2000

top suspect is the actual rumor source divided by the total number of experiments. When the monitor are randomly chosen, the precision of RSD increases from 18% (monitor number: 50) up to 87% (monitor number: 2000), the latter of which is quite effective in practice. This series of experiments also suggest that when the number of monitor is small, in order to identify rumor source accurately, the monitor can be chosen using ID (with precision 45%, monitor number: 50). In contrast, when the number of monitor is large, random is the best way to choose monitors while at the same maintains high accuracy.

3.3 Detecting Multiple Rumor Sources in Networks with Partial Observations

Suppose there are more than one rumor sources in the network; the problem is how to detect all of them based on limited information about network structure and the rumor infected nodes. If each rumor source initiates a different rumor, then the problem can be decoupled to the detection of each rumor source independently. Thus, we assume that one rumor is initiated at a lot of different locations. We place some nodes $v \in K \subseteq V$ as the observers which has a clock that can record the time at which the state of v is changed (e.g., knowing a rumor, being infected or contaminated). Typically, the time when the single source originates is unknown. The monitors/observers can only report the times when they receive the information, but no information about senders (i.e., we do not know who infects whom in epidemic networks or who influences whom in social networks). The information is diffused from the sources to any vertex through shortest paths in the network, i.e., as soon as a vertex receives the information, it sends the information to all its neighbors simultaneously, which takes one time unit. Our goal is to select a subset K of vertices with minimum cardinality such that the source can be uniquely located by the infected times of vertices in K and network structure. This problem is equivalent to finding a minimum set resolving set (SRS) in networks defined in our models.

In this section, we originally propose the concept of set resolving set (SRS) problem and give the formal definition of it. Based on SRS, we then present a novel approach for locating multiple information sources on general graphs with partial observations of the set of infected nodes at the observation time, without knowing the neighbors from which the infection is received. Our method is robust to network heterogeneity and the number of observed infected nodes. To the best of our knowledge, this paper is the first work to study multiple rumor source detection (MRSD) problem via SRS.

Moreover, we show that our objective function for detecting multiple rumor sources in networks is monotone and submodular. By exploiting the submodularity of the objective, we develop an efficient greedy approximation for MRSD problem, which is theoretically proved to have a $(1 + r \ln n + \ln \log_2 \gamma)$-approximation ratio in real world, given that γ is the maximum number of equivalence classes divided by one node-pair. These guarantees are important in practice, since selecting nodes is expensive, and we desire solutions which are not too far from the optimal solution.

The following section is organized as follows. In Sect. 3.3.1, we present the SRS based model and give a formal problem formulation. In Sect. 3.3.2, we then develop a greedy algorithm, and prove its approximation ratio. To confirm the effectiveness of our algorithm, in Sect. 3.3.3, the performance of our algorithm is evaluated in networks which exemplify different structures.

3.3.1   The Model

We start by describing the model and problem statement of multi-rumor-source detection. In the process, we will give the definition of set resolving set (SRS), which is the basis of the model.

If a node u is a rumor source, then we use $s(u)$ to denote the time that it initiates the rumor. If u is not a rumor source, then $s(u) = \infty$. For two nodes u and v, the distance between them is denoted as $d(u, v)$. The time that a rumor initiated at node u is received by node v is $r_u(v) = s(u) + d(u, v)$. For a set of rumor sources $A \subseteq V$, the time that the rumor from A is received by node v is $r_A(v) = \min\{r_u(v) : u \in A\}$.

Definition 3.3 (Set Resolving Set (SRS)) Let K be a node subset of V. Two node set $A, B \subseteq V$ are *distinguishable* by K if there exist two nodes $x, y \in K$ such that

$$r_A(x) - r_A(y) \neq r_B(x) - r_B(y)$$

For a node set $A \subseteq V$, a node $x \in A$ is detectable if A and $A \setminus \{x\}$ are distinguishable by V. Node set A is *detectable* if every node in A is detectable. Let $\mathscr{D}$ be the family of detectable node sets. Node set $K \subseteq V$ is an SRS if any different detectable node sets $A, B \in \mathscr{D}$ are distinguishable by K.

Multi-Rumor-Source Detection problem (MRSD): find an SRS K with the smallest cardinality.

The following theorem characterizes the condition under which a node set is detectable. The idea behind the condition is as follows: when a rumor is initiated at a node x after x can receive the same rumor from some other nodes, then one cannot tell whether the rumor is initiated by x or x merely relays the rumor.

Theorem 3.1 *A node set A is detectable if and only if for every node $x \in A$,*

$$s(x) < r_{A\setminus\{x\}}(x) \tag{3.7}$$

Proof Suppose there is a node $x \in A$ such that $s(x) \geq r_{A\setminus\{x\}}(x)$. Then, there is a node $z \in A \setminus \{x\}$ such that $s(x) \geq r_z(x) = s(z) + d(z, x)$. For any node $y \in V$, $r_x(y) = s(x) + d(x, y) \geq s(z) + d(z, x) + d(x, y) \geq s(z) + d(z, y) = r_z(y)$.

Hence, $r_A(y) = \min\{r_x(y), r_{A\setminus\{x\}}(y)\} = r_{A\setminus\{x\}}(y)$. It follows that $r_A(y_1) - r_A(y_2) = r_{A\setminus\{x\}}(y_1) - r_{A\setminus\{x\}}(y_2)$ for any nodes $y_1, y_2 \in V$, and thus A and $A \setminus \{x\}$ are not distinguishable by V. This finishes the proof for the necessity.

To show the sufficiency, notice that $s(x) < r_{A\setminus\{x\}}(x)$ implies that

$$r_A(x) = \min\{r_x(x), r_{A\setminus\{x\}}(x)\} = \min\{s(x), r_{A\setminus\{x\}}(x)\}$$
$$= s(x) < r_{A\setminus\{x\}}(x) \tag{3.8}$$

For any node $y_1 \in A$, choose $y_2 \in A$ such that $s(y_2) = \min_{y\in A\setminus\{y_1\}}\{s(y)\}$. Then

$$r_{A\setminus\{y_1\}}(y_2) = s(y_2) = r_A(y_2) \tag{3.9}$$

This is because of property (3.8) and the observation $s(y_2) = r_{y_2}(y_2) \geq r_{A\setminus\{y_1\}}(y_2) = \min_{y\in A\setminus\{y_1\}}\{s(y) + d(y, y_2)\} \geq \min_{y\in A\setminus\{y_1\}}\{s(y)\} = s(y_2)$. Also by (3.8), we have

$$r_A(y_1) < r_{A\setminus\{y_1\}}(y_1) \tag{3.10}$$

Combining (3.9) and (3.10), we have

$$r_A(y_1) - r_A(y_2) < r_{A\setminus\{y_1\}}(y_1) - r_{A\setminus\{y_1\}}(y_2)$$

So, A and $A \setminus \{y_1\}$ are distinguishable by y_1 and y_2. The sufficiency follows from the arbitrariness of y_1.

Remark 3.1 If the starting time for all nodes is a constant, then condition (3.7) is satisfied at all nodes. So, this condition does occur in the real world.

Lemma 3.1 *Let A, B be two detectable node sets with $A \setminus B \neq \emptyset$. Then for any node $x \in A \setminus B$ and any node $y \in B$, node sets A and B are distinguishable by x and y.*

Proof Suppose the lemma is not true, then there exists a node $x \in A \setminus B$ and a node $y \in B$ such that

$$r_A(x) - r_A(y) = r_B(x) - r_B(y) \tag{3.11}$$

Since both A and B are detectable, we see from property (3.8) that

$$r_A(x) = s(x) \text{ and } r_B(y) = s(y)$$

Combining these with (3.11), we have

$$s(x) - r_A(y) = r_B(x) - s(y) \le r_y(x) - s(y) = d(y, x) \tag{3.12}$$

Then,

$$r_x(y) = s(x) + d(y, x) \le r_A(y) \le r_x(y) \tag{3.13}$$

It follows that the inequalities in (3.12) and (3.13) become equalities, that is,

$$r_A(y) = r_x(y) \text{ and } r_B(x) = r_y(x)$$

But then,

$$r_A(x) - r_A(y) = s(x) - r_x(y) = -d(x, y) < d(y, x) = r_y(x) - s(y) = r_B(x) - r_B(y)$$

contradicting (3.11). The lemma is proved.

As a consequence of Lemma 3.1, we have the following theorem.

Theorem 3.2 *Node set V is an SRS.*

Theorem 3.2 shows that V is a trivial solution to the MRSD problem. In next section, we shall present an approximation algorithm for the problem.

3.3.2   The Algorithm

In this section, we present a greedy algorithm for MRSD. The algorithm starts from $\mathcal{T} = \emptyset$, and iteratively adds into $\mathcal{T}$ node-pairs with the highest efficiency (which will be defined later) until all sets can be distinguished by some node-pair in $\mathcal{T}$. The output of the algorithm is $K = \bigcup_{T \in \mathcal{T}} T$.

3.3.2.1 Potential Function

The efficiency of a node-pair is related with a potential function f defined as follows. Two detectable node sets A and B are *equivalent* under $\mathcal{T}$, denoted as $A \equiv_{\mathcal{T}} B$, if A and B are not distinguishable by any node-pair in $\mathcal{T}$. Under $\equiv_{\mathcal{T}}$, detectable node sets $\mathcal{F}$ is divided into equivalence classes. The equivalence class containing detectable node set A is denoted as $[A]_{\mathcal{T}}$. Suppose the equivalence classes under $\equiv_{\mathcal{T}}$ are $\mathcal{F}_1, \ldots, \mathcal{F}_k$. Define $\pi(\mathcal{T}) = \prod_{i=1}^{k} |\mathcal{F}_i|!$ and

$$f(\mathcal{T}) = -\log_2 \pi(\mathcal{T}) \tag{3.14}$$

For a node-pair $T = \{x, y\}$, let

$$\Delta_T f(\mathcal{T}) = f(\mathcal{T} \cup \{T\}) - f(\mathcal{T})$$

We shall show that f is monotone increasing and submodular. The proof idea is similar to the one in [17] which studies group testing. The difference is that in [17], only elements need to be distinguished. While in this paper, distinguishing sets needs more technical details. The following is a technical result of combinatorics (see Fig. 3.6a for an illustration of its conditions).

Lemma 3.2 *Suppose* $\{h_{ij}\}_{i=1,\ldots,p}^{j=1,\ldots,q}$ *is a set of non-negative integers. For* $i = 1, \ldots, p$, $a_i = \sum_{j=1}^{q} h_{ij}$. *For* $j = 1, \ldots, q$, $b_j = \sum_{i=1}^{p} h_{ij}$. *Furthermore,* $\sum_{i=1}^{p} a_i = \sum_{j=1}^{q} b_j = g$. *Then*

$$\frac{g!}{\prod_{j=1}^{q} b_j!} \geq \frac{\prod_{i=1}^{p} a_i!}{\prod_{i=1}^{p} \prod_{j=1}^{q} h_{ij}!} \tag{3.15}$$

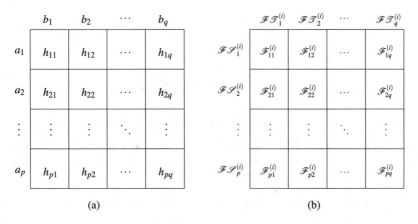

(a)   (b)

Fig. 3.6 (a) Illustration for the conditions in Lemma 3.2. (b) Refinement of equivalence class $\mathcal{F}_i$ by adding S and T

Proof Consider the expansion of the following multi-variable polynomial:

$$(x_{11} + \cdots + x_{1q})^{a_1} \cdots (x_{p1} + \cdots + x_{pq})^{a_p}$$

$$= \left(\sum_{a_{11}+\cdots+a_{1q}=a_1} \frac{a_1!}{\prod_{j=1}^{q} a_{1j}!} x_{11}^{a_{11}} \cdots x_{1q}^{a_{1q}} \right) \cdots$$

$$\left(\sum_{a_{p1}+\cdots+a_{pq}=a_p} \frac{a_p!}{\prod_{j=1}^{q} a_{pj}!} x_{p1}^{a_{p1}} \cdots x_{pq}^{a_{pq}} \right)$$

$$= \sum \frac{\prod_{i=1}^{p} a_i!}{\prod_{i=1}^{p} \prod_{j=1}^{q} a_{ij}!} x_{11}^{a_{11}} \cdots x_{1q}^{a_{1q}} \cdots x_{p1}^{a_{p1}} \cdots x_{pq}^{a_{pq}}$$

where the sum is over all non-negative integers $\{a_{ij}\}_{i=1,\ldots,p}^{j=1,\ldots,q}$ satisfying $\sum_{j=1}^{q} a_{ij} = a_i$ for $i = 1, \ldots, p$. Setting $x_{1j} = \cdots = x_{pj} = x_j$ for $j = 1, \ldots, q$ in the above equation, we have

$$(x_1 + \cdots x_q)^{a_1 + \cdots + a_p}$$

$$= \sum \frac{\prod_{i=1}^{p} a_i!}{\prod_{i=1}^{p} \prod_{j=1}^{q} a_{ij}!} x_1^{a_{11}+\cdots+a_{p1}} \cdots x_q^{a_{1q}+\cdots+a_{pq}} \qquad (3.16)$$

On the other hand,

$$(x_1 + \cdots x_q)^{a_1 + \cdots + a_p}$$

$$= (x_1 + \cdots x_q)^g = \sum_{r_1+\cdots r_q=g} \frac{g!}{\prod_{j=1}^{q} r_j!} x_1^{r_1} \cdots x_q^{r_q} \qquad (3.17)$$

Comparing the coefficients of $x_1^{b_1} \cdots x_q^{b_q}$ in (3.16) and (3.17), we have

$$\frac{g!}{\prod_{j=1}^{q} b_j!} = \sum \frac{\prod_{i=1}^{q} a_p!}{\prod_{i=1}^{p} \prod_{j=1}^{q} a_{ij}!} \qquad (3.18)$$

where the sum is over all non-negative integers $\{a_{ij}\}_{i=1,\ldots,p}^{j=1,\ldots,q}$ satisfying $\sum_{j=1}^{q} a_{ij} = a_i$ for $i = 1, \ldots, p$ and $\sum_{i=1}^{p} a_{ij} = b_j$ for $j = 1, \ldots, q$. Since $\{h_{ij}\}_{i=1,\ldots,p}^{j=1,\ldots,q}$ satisfy these restrictions, the right-hand side of (3.15) is one term contained in the right-hand side of (3.18). Then, the Lemma follows.

We shall use the following characterization of monotonicity and submodularity.

Lemma 3.3 ([55, Lemma 2.25]) *Let f be a function defined on all subsets of a set U. Then f is submodular and monotone increasing if and only if for any two subsets $R \subseteq S \subseteq U$ and any element $x \in U$,*

$$\Delta_x f(R) \geq \Delta_x f(S)$$

Lemma 3.4 *The function f defined in (3.14) is submodular and monotone increasing.*

Proof To use Lemma 3.3, we are to show that for any families of node-pairs $\mathscr{T}_1 \subseteq \mathscr{T}_2$ and any node-pair T,

$$\Delta_T f(\mathscr{T}_1) \geq \Delta_T f(\mathscr{T}_2) \tag{3.19}$$

In fact, it suffices to prove (3.19) for the case that $|\mathscr{T}_2 \setminus \mathscr{T}_1| = 1$. Then, induction argument will yield the result for the general case. So, in the following, we assume that $\mathscr{T}_2 = \mathscr{T}_1 \cup \{S\}$, where S is a node-pair. In this case, (3.19) is equivalent to

$$\frac{\pi(\mathscr{T}_1)}{\pi(\mathscr{T}_1 \cup \{T\})} \geq \frac{\pi(\mathscr{T}_1 \cup \{S\})}{\pi(\mathscr{T}_1 \cup \{S, T\})} \tag{3.20}$$

Suppose the equivalence classes under $\equiv_{\mathscr{T}}$ are $\mathscr{F}_1, \ldots, \mathscr{F}_k$. Notice that for any detectable node set A, $[A]_{\mathscr{T} \cup \{S\}} \subseteq [A]_{\mathscr{T}}$, that is, adding one node-pair results in a refinement of equivalence classes. Also notice that for any detectable node set A,

$$[A]_{\mathscr{T} \cup \{S, T\}} = [A]_{\mathscr{T} \cup \{S\}} \cap [A]_{\mathscr{T} \cup \{T\}}$$

Hence we may assume (see Fig. 3.6b for an illustration) that for each $i = 1, \ldots, k$,

(a) equivalence classes under $\mathscr{T} \cup \{S, T\}$ which are contained in $\mathscr{F}_i$ are $\{\mathscr{F}_{s,t}^{(i)}\}_{s=1,\ldots,l_i}^{t=1,\ldots,m_i}$;

(b) For $s = 1, \ldots, l_i$, let $\mathscr{FS}_s^{(i)} = \bigcup_{t=1}^{m_i} \mathscr{F}_{s,t}^{(i)}$. Equivalence classes under $\equiv_{\mathscr{T} \cup \{S\}}$ contained in $\mathscr{F}_i$ are $\{\mathscr{FS}_s^{(i)}\}_{s=1}^{l_i}$;

(c) For $t = 1, \ldots, m_i$, let $\mathscr{FT}_t^{(i)} = \bigcup_{s=1}^{l_i} \mathscr{F}_{s,t}^{(i)}$. Equivalence classes under $\equiv_{\mathscr{T} \cup \{T\}}$ contained in $\mathscr{F}_i$ are $\{\mathscr{FT}_t^{(i)}\}_{t=1}^{m_i}$.

Taking $a_l = |\mathscr{FS}_l^{(i)}|$, $b_j = |\mathscr{FT}_j^{(i)}|$, $h_{lj} = |\mathscr{F}_{lj}^{(i)}|$, and $g = |\mathscr{F}_i|$, the conditions of Lemma 3.2 are satisfied, and thus

$$\frac{|\mathscr{F}_i|!}{\prod_{j=1}^{q} |\mathscr{FT}_j^{(i)}|!} \geq \frac{\prod_{l=1}^{p} |\mathscr{FS}_l^{(i)}|!}{\prod_{l=1}^{p} \prod_{j=1}^{q} |\mathscr{F}_{lj}^{(i)}|!}$$

which is exactly the desired inequality (3.20).

Lemma 3.5 *Suppose node-pair T divides $\mathscr{F}$ into k equivalence classes. Then*

$$\Delta_T f(\emptyset) \leq |\mathscr{F}| \log_2 k$$

Proof Suppose the equivalence classes under $\equiv_T$ have cardinalities $n_1, \ldots, n_k$, respectively. Then

$$\Delta_T f(\emptyset) = f(\{T\}) - f(\emptyset) = \log_2 \left(\frac{|\mathscr{F}|!}{\prod_{i=1}^k n_i!} \right) \leq \log_2 \left(k^{|\mathscr{F}|} \right) = |\mathscr{F}| \log_2 k$$

where the inequality can be seen by setting $x_1 = \cdots = x_k = 1$ in the following equation:

$$(x_1 + \cdots + x_k)^{|\mathscr{F}|} = \sum_{n_1 + \cdots + n_k = n} \frac{|\mathscr{F}|!}{\prod_{i=1}^k n_i!} x_1^{n_1} \ldots x_k^{n_k}$$

The lemma is proved.

3.3.2.2   The Algorithm and Its Approximation Ratio

As stated at the beginning of Sect. 3.3.2, an SRS will be derived from a family $\mathscr{T}$ of node-pairs such that

$$\text{node-pairs in } \mathscr{T} \text{ can distinguish all detectable node sets.} \tag{3.21}$$

Call any family of node-pairs as a *test family*, and call a test family satisfying condition (3.21) as a *valid test family*.

Lemma 3.6 *Suppose $\mathscr{T}$ is a valid test family. Let $K = \bigcup_{T \in \mathscr{T}} T$ and x be an arbitrary node in K. Then $\widetilde{\mathscr{T}} = \{(x, y): y \in K \setminus \{x\}\}$ is also a valid text family.*

Proof Observe that if two detectable node sets A, B are distinguished by $\{y, z\} \in \mathscr{T}$, then they can be distinguished by either $\{x, y\}$ or $\{x, z\}$. The lemma follows.

Notice that all node-pairs in $\widetilde{\mathscr{T}}$ have a common element. We call such a test family as a *canonical test family*. Notice that $\bigcup_{T \in \mathscr{T}} T = \bigcup_{T \in \widetilde{\mathscr{T}}} T = K$. Hence $\mathscr{T}$ and $\widetilde{\mathscr{T}}$ are equivalent in the sense that they produce a same SRS. As a consequence, to find an SRS, it suffices to consider canonical test families, that is, to find a node x and a valid test family $\mathscr{T}_x \subseteq \mathscr{P}_x = \{\{x, y\}: y \in V \setminus \{x\}\}$.

In order to analyze the approximation ratio, we have to compare the size of the approximation solution with that of an optimal one. Since we do not know which node is in an optimal solution, we have to "guess." To be more concrete, for each node $x \in V$, the algorithm finds a valid test family $\mathscr{T}_x \subseteq \mathscr{P}_x$. Let $K_x = \bigcup_{T \in \mathscr{T}_x} T$.

Algorithm 3 Greedy algorithm for MRSD

Input: A graph $G = (V, E)$.
Output: A node set K which is an SRS.
 for all $x \in V$ **do**
  Set $\mathcal{T}_x \leftarrow \emptyset$.
  while there exists a node-pair $T \in \mathcal{P}_x$ such that $\Delta_T f(\mathcal{T}_x) > 0$ **do**
   select node-pair $T \in \mathcal{P}_x$ with the maximum $\Delta_T f(\mathcal{T}_x)$.
   $\mathcal{T}_x \leftarrow \mathcal{T}_x \cup \{T\}$.
  end while
  $K_x = \bigcup_{T \in \mathcal{T}_x} T$.
 end for
 Output $K \leftarrow \arg\min\{|K_x| : x \in V\}$.

The final output of the algorithm is $K = \arg\min_{x \in V} |K_x|$. The details of the algorithm for MRSD is described in Algorithm 3.

Lemma 3.7 *A test family $\mathcal{T}$ is valid if and only if $\Delta_T f(\mathcal{T}) = 0$ for any node-pair T.*

Proof First, we make some observation. Suppose the equivalence classes under $\mathcal{T}$ are $\mathcal{F}_1, \ldots, \mathcal{F}_k$. For each $i = 1, \ldots, k$, $\mathcal{F}_i$ is refined under $\mathcal{T} \cup \{T\}$ into equivalence classes $\mathcal{F}_1^{(i)}, \ldots, \mathcal{F}_{l_i}^{(i)}$. Then

$$
\begin{aligned}
\Delta_T f(\mathcal{T}) &= \log_2 \left(\frac{\prod_{i=1}^{k} |\mathcal{F}_i|}{\prod_{i=1}^{k} \prod_{j=1}^{l_i} |\mathcal{F}_j^{(i)}|} \right) \\
&= \log_2 \left(\prod_{i=1}^{k} \frac{|\mathcal{F}_i|}{\prod_{j=1}^{l_i} |\mathcal{F}_j^{(i)}|} \right)
\end{aligned}
\tag{3.22}
$$

Notice that $|\mathcal{F}_i| / \prod_{j=1}^{l_i} |\mathcal{F}_j^{(i)}|$ is the number of ways to put $|\mathcal{F}_i|$ balls into l_i labeled boxes such that the j-th box contains $|\mathcal{F}_j^{(i)}|$ balls ($j = 1, \ldots, l_i$). So,

$$
|\mathcal{F}_i| / \prod_{j=1}^{l_i} |\mathcal{F}_j^{(i)}|
\tag{3.23}
$$

is a positive integer which equals 1 if and only if $l_i = 1$.

Notice that $l_i = 1$ implies that adding node-pair T into $\mathcal{T}$ does not incur a strict refinement of $\mathcal{F}_i$.

If $\mathcal{T}$ is a valid test family, then every equivalence class has cardinality 1, and thus $f(\mathcal{T}) = 0$. Combining this with the fact that f is a non-positive monotone increasing function, we see that the maximum value of f is zero and thus $\Delta_T(\mathcal{T}) = 0$ holds for any node-pair T.

If $\mathcal{T}$ is not a valid test family, then there exist two different detectable node sets A, B which cannot be distinguished by $\mathcal{T}$. Since $A \neq B$, we may assume that $A \setminus B \neq \emptyset$. By Lemma 3.1, A, B can be distinguished by a node-pair $\{y, z\}$ with $y \in A \setminus B$ and $z \in B$. Then by Lemma 3.6, A, B can be distinguished by $T = \{x, y\}$ or $\{x, z\}$. In this case, at least one equivalence class under $\mathcal{T}$ is refined by adding T. In other words, there is an $i \in \{1, \ldots, k\}$ such that $|\mathcal{F}_i| / \prod_{j=1}^{l_i} |\mathcal{F}_j^{(i)}| > 1$. Then by (3.22), $\Delta_T(\mathcal{T}) > 0$. The lemma is proved.

Theorem 3.3 *Suppose γ is the maximum number of equivalence classes divided by one node-pair. Then, Algorithm 3 correctly computes an SRS with approximation ratio at most $1 + \ln\left(|\mathcal{F}| \log_2 \gamma\right)$.*

Proof By Lemma 3.7, we see that every $\mathcal{T}_x$ computed in the algorithm is a valid test family. The correctness follows.

To analyze the approximation ratio, suppose K^* is an optimal solution to MRSD and x is a node in K^*. Let $\mathcal{T}^* = \{\{x, y\}: y \in K^* \setminus \{x\}\}$.

Consider the test family $\mathcal{T}_x$ produced by the greedy algorithm for node x. We claim that every node-pair T chosen in the algorithm satisfies

$$\Delta_T f(\mathcal{T}_x) \geq 1 \tag{3.24}$$

By expression (3.22), this is equivalent to show

$$\prod_{i=1}^{k} \frac{|\mathcal{F}_i|}{\prod_{j=1}^{l_i} |\mathcal{F}_j^{(i)}|} \geq 2 \tag{3.25}$$

Since every T taken in the algorithm has $\Delta_T f(\mathcal{T}_x) > 0$, which is equivalent to $\prod_{i=1}^{k} \frac{|\mathcal{F}_i|}{\prod_{j=1}^{l_i} |\mathcal{F}_j^{(i)}|} > 1$, we see that at least one $|\mathcal{F}_i| / \prod_{j=1}^{l_i} |\mathcal{F}_j^{(i)}|$ is greater than 1, and thus is at least 2. Inequality (3.25) follows from this observation and property (3.23). Claim (3.24) is proved.

We shall use Theorem 3.7 in [55], which says, using terminologies here, that as long as (3.24) is true, then

$$|\mathcal{T}_x| \leq \left(1 + \ln \frac{f(\mathcal{T}^*) - f(\emptyset)}{|\mathcal{T}^*|}\right) \cdot |\mathcal{T}^*| \tag{3.26}$$

By a property of submodular function (see [55, Lemma 2.23]),

$$\sum_{T \in \mathcal{T}^*} \Delta_T f(\emptyset) \geq \Delta_{\mathcal{T}^*} f(\emptyset) = f(\mathcal{T}^*) - f(\emptyset)$$

Combining this with Lemma 3.5,

$$
\begin{aligned}
&\frac{f(\mathscr{T}^*) - f(\emptyset)}{|\mathscr{T}^*|} \\
&\le \frac{\sum_{T \in \mathscr{T}^*} \Delta_T f(\emptyset)}{|\mathscr{T}^*|} \le \max_{T \in \mathscr{T}^*} \{\Delta_T f(\emptyset)\} \le |\mathscr{F}| \log_2 \gamma
\end{aligned}
\tag{3.27}
$$

Combining (3.26), (3.27) with $|K_x| = |\mathscr{T}_x| - 1$ and $|K^*| = |\mathscr{T}^*| - 1$, we have

$$
|K_x| \le \left(1 + \ln\left(|\mathscr{F}| \log_2 \gamma\right)\right)\left(|K^*| - 1\right) + 1 \le \left(1 + \ln\left(|\mathscr{F}| \log_2 \gamma\right)\right)|K^*|
$$

The approximation ratio follows since the algorithm chooses a node x_0 with $|K_{x_0}| = \min_{y \in V} |K_y| \le |K_x|$.

Remark 3.2 Notice that in a worst case, the number of detectable node sets is $\Theta(2^n)$. In fact, if the starting time for all nodes is a constant, then by Remark 3.1, every nonempty node set is detectable, and thus $|\mathscr{F}| = 2^n - 1$. In such a case, the approximation ratio is $(1 + n \ln 2 + \ln \log_2 \gamma)$, which is no better than a trivial bound n. However, in the real world, it is reasonable to assume that the number of rumor sources is at most a constant number r, and only those detectable node sets of cardinality at most r are considered. In this case, $|\mathscr{F}| = O(n^r)$, and the approximation ratio is $(1 + r \ln n + \ln \log_2 \gamma)$.

Remark 3.3 Notice that $\gamma \le 2D + 1$, where D is the diameter of the graph. To see this, suppose $T = \{x, y\}$ is a node-pair, A is a node set, and $r_A(x) - r_A(y) = c$, then a node set B belongs to equivalence class $[A]_{\{T\}}$ if and only if $r_B(x) - r_B(y) = c$. Notice that c has at most $2D + 1$ different values, namely, $\{-D, -(D - 1), \ldots, -1, 0, 1, \ldots, D - 1, D\}$. So, one node-pair divides $\mathscr{F}$ into at most $2D + 1$ equivalence classes. In a social network, D is a small constant. So, the third term $\ln \log_2 \gamma$ in the above approximation ratio is not large.

3.3.3   Simulation Results

In this section we experimentally evaluate our greedy algorithm for MRSD, in particular its effectiveness in finding rumor sources—how many sources it identifies, whether it correctly identifies and its scalability.

As discussed in the introduction, the existing proposals for identifying rumor sources consider significantly different problems settings than we do. The rumor centrality of Shah and Zaman [171, 173] can only discover one rumor source, while estimators proposed in [123] consider a completely different infection model from ours. As such it is not meaningful to compare performances, and therefore here we only consider the greedy algorithm for MRSD.

In our study we conduct simulations on synthetic networks exemplifying different types of structure—including geometric trees, regular trees, and small-world networks. In general, we set the rumor sources, simulate diffusion process, and record the times of monitors when they received the rumors. Then, given the times and network structure, we try to infer the number rumor sources and where they are.

As described in Sect. 3.3.1, the diffusion model is implemented as a discrete event in Java. Each hop takes one time unit. Note that the cascade starts from all rumor sources at the same time stamp. The number of rumor sources is set as k. For each k, we perform large number of simulation runs to get high precession.

3.3.3.1   Effectiveness of Greedy in Identifying How Many

The number of infection sources k are chosen to be 1, 2, 3, and 4. For each type of network and each number of infection sources, we perform 1000 simulation runs with 500 monitors. The estimation results for the number of infection sources in different scenarios are shown in Fig. 3.7. It can be seen that our algorithm correctly finds the number of infection sources more than 95% of the time for geometric trees, and more than 86% of the time for regular trees. The accuracy of about 79% for small-world networks is slightly lower than that for the tree networks, as the node-pair for a small-world network is estimated based on the BFS heuristics, thus additional errors are introduced into the procedure. It also shows the power of our approach, as we can easily identify the true number of seeds for most cases using a principled approach.

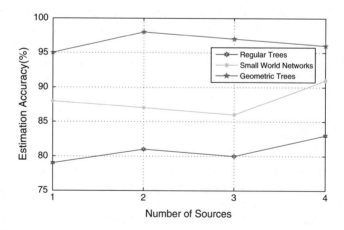

Fig. 3.7 Estimating the number of rumor sources

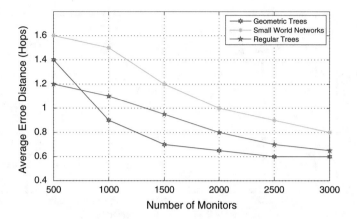

Fig. 3.8 Estimating the average error distance between the identified source and the actual source

3.3.3.2   Effectiveness of Greedy in Identifying Which Ones

To quantify the performance of identified rumor sources, we propose error distance. Error distance is defined as the average distances between the estimates and the respective rumor sources. To be specific, we match the estimated source nodes with the actual sources so that the sum of the error distances between each estimated source and its match is minimized. If we have incorrectly estimated the number of infection sources, we neglect the extra number of found nodes since here we only focus on the error distance between correct sources. In Fig. 3.8, we see that the proposed algorithm finds rumor sources that have small error distance on average. Note that the reported results here are also based on 1000 trials. For geometric trees, the average error distance lies between 1.4 and 0.6 hop. For regular trees, the error distance decreases from 1.2 to 0.65 hops. For small-world networks, the value is between 1.6 and 0.8. In general, the average error distance is less than two hops. Moreover, as the number of monitors increases, error distance will start to drop.

3.3.3.3   Scalability

Figure 3.9 demonstrates the average computation time of greedy after running it on increasingly larger infected graphs (as the complexity depends on the size of the monitors). We use small-world network graph with $k = 2$. The statistics of running time is based on 10 runs for each graph. As we can see, the running time is linear on the number of edges of the infected graph. Thus, overall our algorithm scales well with high performance in solution quality.

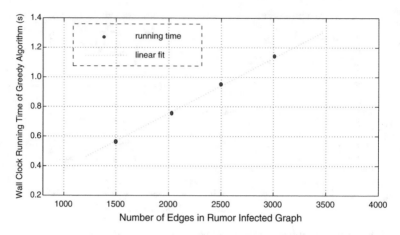

Fig. 3.9 Wall-clock computation time (in seconds) by our greedy algorithm for increasingly larger infected graphs. $k = 2$. Each point average of 10 runs

3.4   Conclusion

In this chapter, we have studied the rumor source detection problem under different scenarios. In Sect. 3.2, we have investigated the problem of rumor source detection in online social networks when lacking of text information. We formalize the problem and define rumor quantifier, a probability based score for ranking how likely a node is going to be the actual rumor source. The idea behind it is simple: a cascade is more likely to spread from a rumor source to the monitors who have received the information but less likely to those who have not. To compute the rumor quantifier of each node, we developed a scalable algorithm, RSD, to detect the rumor source and differentiate the rumors.

We evaluated various monitor deployment on real online social network—Twitter—with rumor propagating according to the popular independent cascade model, and showed that our algorithm, RSD is able to accurately identify the rumor source from a large network when there are reasonable number of monitors. Our future work includes two aspects: first, discuss the situation when there are multiple rumor sources spreading in the network. Second, we will try to develop both structure and content combined method to identify rumor sources.

In Sect. 3.3 we discussed finding multiple rumor sources, the challenging problem of identifying the nodes from which an infection in a graph started to spread. We first gave the definition of set resolving set(SRS) and proposed to employ minimum SRS for identifying the set of rumor sources from which the rest of nodes in the graph can be distinguished correctly. In this framework, the inference is based only on knowledge of the infected monitors and the underlying network structure. We have designed a highly efficient greedy algorithm using submodularity analysis and theoretically proved the performance ratio to be $1 + \ln\left(|\mathscr{F}|\log_2 \gamma\right)$, given that γ is the maximum number of equivalence classes divided by one node-pair.

Several improvements and future directions are possible. One direction is to extend our methodology to different applications, including influence maximization, rumor blocking, etc. and see how the proposed methodology leads to deeper insights. Another promising direction is to tackle the MRSD problem in different diffusion models, such as models with transmission probabilities between nodes considered or models without submodularity property.

Chapter 4
Rumor Blocking in Social Networks

4.1 Overview

Online social networks have many benefits as a medium for fast, widespread information dissemination. They provide fast access to large scale news data, sometimes even before the mass media. They also serve as a medium to collectively achieve a social goal. For instance with the use of group and event pages in Facebook, events such as Day of Action protests reached thousands of protestors [71]. While the ease of information propagation in social networks can be very beneficial, it can also have disruptive effects. One such example was observed in August, 2012, thousands of people in Ghazni province left their houses in the middle of the night in panic after the rumor of earthquake [127]. Another example is the fast spread of misinformation in twitter that the president of Syria is dead, leading to a sharp, quick increase in the price of oil [208]. There are lots of similar examples. Although social networks are the main source of news for many people today, they are not considered reliable due to such problems.

Clearly, in order for social networks to serve as a reliable platform for disseminating critical information, it is necessary to have tools to limit the effect of misinformation or rumors. Existing work in controlling rumor spread includes [30, 62, 63, 84, 98, 147]. In [98], Kimura et al. proposed to block a certain number of links in a network to reduce the bad effects of rumors. In the presence of a misinformation cascade, [30, 62, 63, 84, 147] aim to find a near-optimal way of disseminating good information that will minimize the devastating effects of a misinformation campaign. For instance, [84] seeks ways of making sure that most of the users of the social network hear about the correct information before the bad one, making social networks a more trustworthy or reliable source of information.

Related Work The identification of influential users in a social network is a problem that has received significant attention in recent research. For the influence maximization problem, given a probabilistic model of information diffusion such

© The Author(s), under exclusive license to Springer Nature Switzerland AG 2020
W. Xu, W. Wu, *Optimal Social Influence*, SpringerBriefs in Optimization,
https://doi.org/10.1007/978-3-030-37775-5_4

as the independent cascade model, a network graph, and a budget k, the objective is to select a set S of size k for initial activation so that the expected value of $f(S)$ (size of cascade created by selecting seed set A) is maximized [37–39, 96, 97, 122, 139, 197, 207]. For learning more about this topic, please refer to our related book chapter of influence maximization.

In contrast to influence maximization problem which studies single-cascade influence propagation (only one kind of influence diffuses in a social network), there is a series of work that focus on multiple-cascade influence diffusion in social networks. Bharathi et al. [19] explored the multiple-cascade influence diffusion under the extension of the independent cascade (IC) model. In [26], Borodin et al. studied the multiple-cascade influence diffusion in several different models generated from the linear threshold (LT) model. In [190], Trpevski et al. proposed a two-cascade influence diffusion model based on the SIS (susceptible-infected-susceptible) model. Kostka et al. [104] considered the two-cascade influence diffusion problem from a game-theory aspect, where each cascade tries to maximize their influence among the social network. Then they studied it under a more restricted model than the IC model and the LT model. To learn more about the study in game-theoretic models where multiple decision-makers try to maximize their own objectives at the same time, interested readers are referred to [5, 33, 147, 191, 192, 195].

Among multiple-cascade influence diffusion, there is a special kind, rumor control related problem, in which there are only two kinds of cascades, one is called positive cascade, while the other is called negative cascade. The goal is to use the positive cascade diffusion to fight against the negative cascade diffusion. Budak et al. [30] and He et al. [84] focused on the problem: given a set of initial "bad" seeds, how to optimally choose the initial set of "good" seeds to limit the diffusion of their influence? In [30], the authors proved the NP-hardness of this optimization problem under the generalized IC model. They also established the submodularity of the objective function and therefore, the greedy algorithm was used as a constant-factor approximation algorithm. In [84], He et al. proposed a competitive linear threshold (CLT) model. They proved that the objective function is submodular and obtained a $(1-1/e)$-approximation ratio. To overcome the inefficiency of the greedy algorithm, they designed a heuristic algorithm which uses the local structure of the network.

Extending both the IC and LT models to two-cascade information diffusion model with a time deadline, Nguyen et al. in [147] studied the following problem: given bad influence sources, how to select the least number of nodes as good influence sources to limit bad influence propagation in the entire network, such that after T steps, the expected number of infected nodes is at most $1 - \beta$. The authors demonstrated several hardness results and proposed effective greedy algorithms and heuristic algorithms.

In this chapter, we will introduce two recent works about rumor blocking or rumor control in detail, including *Community-Based Least Cost Rumor Blocking* (Sect. 4.2) [63] and *Rumor Blocking Maximization with Constrained Time* (Sect. 4.3) [62].

To efficiently decontaminate the wide spread of rumors in a network, in Sect. 4.2, attention is drawn to exploiting communities, i.e., confine the rumor diffusion to its own community. In this work, we propose to initiate protectors to fight against rumors in social networks. That is, we select some individuals as initial protectors, let them spread true or credible information. Correspondingly, some individuals will be protected from rumors. In specific, we focus on protecting bridge ends using certain number of protectors. Here, bridge ends are boundary nodes of communities, which have relations with members in rumor community, and can be reached by rumors at an earlier stage.

In Sect. 4.3, we investigate the problem: given the number of initial protectors and deadline, how to select initial protectors such that the number of "really" protected members in social networks is maximized within deadline. We propose two models to capture competitive influence diffusion process, namely the Rumor-Protector Independent Cascade model with Meeting events (RPIC-M) model and the Rumor-Protector Linear Threshold model with Meeting events (RPLT-M) model. Three features are included in these two models: a time deadline, random time delay between information exchange, and personal interests regarding the acceptance of information. Under these two models, we study the Rumor Containment maximization with the constraints: time Deadline, Meeting events, and Personal interests (RC-DMP problem). We prove that the problem under these two models is both NP-hard. Moreover, we demonstrate that the objective functions for the problem under the two different models are both monotone and submodular. Therefore, we apply the greedy algorithm as a constant-factor approximation algorithm with performance guarantee ratio of $1 - \frac{1}{e}$.

In the last section, we will summarize our work of rumor blocking and future work in this field will also be discussed.

4.2 Community-Based Least Cost Rumor Blocking

We assume that rumors and protectors start diffusing at the same time, and also follow a same diffusion mechanism. Each node can only be in one of the three statuses: *protected*, *infected*, or *inactive*. When the two cascades, namely *cascade P* for protector and *cascade R* for rumor, arrive a node at the same time, we say that cascade P has priority over cascade R, in other words, the node is protected. By considering the community property of a social network, we identify certain kinds of nodes, which are located in boundaries of communities, as protection targets. Then the goal of the *Rumor Control (RC)* problem is to find the minimal number of initial protectors to protect certain fraction of these nodes. This is a novel perspective in constraining rumor dissemination.

The authors in [21] found that a social network is composed of a set of disjoint communities, and members in a same community have similar interests. Furthermore, we have the common knowledge that most of the time, rumors originate from individuals with similar interests. Therefore, we assume that rumors originate from a same community of a social network. According to the community property that connections among individuals in a same community are denser than that across different communities, we know that influence spreads faster within a same community, while slower across different communities.

To simplify the description, we name the community that contains rumors as *rumor community* and a neighbor community of rumor community as a *R-neighbor community*. Considering the number of nodes that we can protect and the number of nodes that we need to use as initial protectors, it is practical for us to protect the members in R-neighbor communities. We focus on those nodes that exist in R-neighbor communities, and also can be reached first when cascade R arrives in their own communities. We name them as *bridge ends*.

We study the RC problem under an influence diffusion model: the *Deterministic One-Activate-Many* (DOAM) model. Considering the budget for launching the initial protectors, in the DOAM model, we focus on the *RC-D* problem, where we need to protect all the bridge ends.

Through proving the equivalence between the RC-D problem and the Set Cover (SC) problem, we propose the *Set Cover Based Greedy* (SCBG) algorithm. Then we demonstrate that there is no polynomial time $o(\ln n)$-approximation for the RC-D problem unless $P = NP$, and get a $O(\ln n)$-approximation ratio solution. Finally, we collect real-world data to validate our algorithms, and the experimental reports demonstrate that both the Greedy algorithm and the SCBG algorithm outperform the other two heuristics, respectively.

The rest of this section is organized as follows: in Sect. 4.2.1, we propose an influence diffusion models, namely, the DOAM model. In Sect. 4.2.2, we formulate the RC-D problem under the proposed model. In Sect. 4.2.3, as for the DOAM model, we prove that there is no polynomial time $o(\ln n)$-approximation for the RC-D problem unless $P = NP$, and propose the SCBG algorithm. In Sect. 4.2.4, we compare our algorithms with other heuristics and analyze the experimental results.

4.2.1 Deterministic One-Activate-Many (DOAM) Propagation Model

A social network can be modeled as a directed graph $G = (V, E)$, in which V denotes the node set and E denotes the edge set. In the context of influence diffusion, V represents the individuals in this network and E represents the relationships among these individuals. Furthermore, a node $u \in V$ is an in-neighbor of a node $v \in V$ if there exists an edge $e_{uv} \in E$ (i.e., the edge from u to v exists in graph G). A node v is called an out-neighbor of u if u is an in-neighbor of v. Based on this special structure, influence can diffuse among individuals in social networks. Since under different situations, influence spreads with different mechanisms. In our paper, we introduce a new influence diffusion model.

Here, we first introduce some denotations and three properties of the model. Let R represent rumor cascade and P denote protector cascade. A node is said to be *infected* (*protected*) if it is influenced by *rumors* (*protectors*) either initially or sequentially from one of its neighbors, or *inactive* otherwise. We also denote the initial set of infected (protected) nodes for R (P) as S_r (S_p).

Right now, we introduce three properties of the proposed model: (1) There are two kinds of cascades R (for rumor) and P (for protector); (2) when R and P reach a node u at the same time, P has the priority to influence u, meaning u will always be protected; (3) R or P diffuses *progressively*, that is, nodes can switch from inactive to infected or protected, but cannot switch in the other direction, that is, once an inactive node is infected or protected, it will never change its status. Property (1) makes sense since it happens in reality. Property (2) is reasonable since people are likely to believe in the truth. Property (3) originates from [93].

In the following, we describe the proposed model in detail.

Given the initial rumor set S_r, an initial protector set S_p is selected and protected at step $t = 0$. At any step $t \geq 1$, when a node u first becomes infected (protected), it will infect (protect) all of its currently inactive out-neighbors successfully. And u only has one chance to influence its out-neighbors, that is, at step $t + 1$, u will not influence its out-neighbors. This influence diffuses in discrete time and continues until no new inactive nodes become protected or infected.

This influence propagation process is actually the information broadcast (one-to-many) phenomenon in social networks, under which situation, each person is able to spread the information to many persons simultaneously. Obviously, the information diffusion speed under the DOAM model is very fast.

4.2.2 Rumor Control Problem

In this section, we define the Rumor Control (RC) problem in social networks. It is known that a social network is composed of individuals and connections between individuals. We notice that social networks have community property, that is, they divide into groups of members, where connections within the same group are dense while across different groups are sparse. It is common sense that individuals form communities based on their common interests, and they are more likely to communicate with members in their own communities than with members in other communities. Therefore, the connections within the same communities are dense while across different communities are sparse. Thus, it is impossible that information can spread fast from one community to other communities.

Based on the community property, to efficiently control the wide spread of rumors originated from one community, we try to prevent them from spreading out to other communities. To realize it, we only need to protect all the members in R-neighbor communities. Bridge ends are the nodes that exist in R-neighbor communities and can be reached first when cascade R arrives in their own communities. Therefore, to protect all the members in R-neighbor communities, it is enough to protect all the bridge ends.

In the following, we give some problem-related definitions and the formal definitions of our problems.

Definition 4.1 A social network is a directed graph $G(V, E, C)$, where each node $v_i \in V$ denotes an individual in the network, and a directed edge $(v_i, v_j) \in E$ denotes the event that individual v_i has influence on individual v_j. Here $C = \{C_1, C_2, \cdots, C_k\}$ is a set of disjoint communities that form the network, satisfying $\bigcup_{r=1}^{k} V(C_r) = V$, where $V(C_r)$ denotes the individuals in community C_r.

Definition 4.2 Rumor Control (RC) problem: Given a community C_k in $G(V, E, C)$, an initial rumor set $S_r \subseteq V(C_k)$ ($C_k \in C$ is the rumor community and is predetermined), and bridge ends B, our goal is to select a least number of nodes as the initial protectors, such that at least α ($0 \leq \alpha \leq 1$) fraction of the bridge ends are protected in the end of influence diffusion.

Considering the influence propagation speed under the DOAM model, we introduce the RC-D problem for the DOAM model. It is because under the DOAM model, rumors propagate very fast in a social network. In other words, within short time, rumors can infect a large amount of individuals in a social network. Considering the budget in launching the initial protectors, the goal of the problem requires to protect all the bridge ends.

Definition 4.3 The RC-D problem: Given a community C_k in $G(V, E, C)$, an initial rumor set $S_r \subseteq V(C_k)$ ($C_k \in C$ is the rumor community and is predetermined), and bridge ends B, under the DOAM model, our goal is to select a least number of nodes as the initial protectors, such that all the bridge ends ($\alpha = 1$ in the RC problem.) are protected in the end of influence diffusion.

Since the *Set Cover (SC)* problem will be used in the RC-D problem, here we give its definition below.

Definition 4.4 Set Cover (SC) Problem: Given a set of elements $U = \{v_1, v_2, \cdots, v_n\}$ and a set of m subsets of U, called $S = \{S_1, S_2, \cdots, S_m\}$, find a "least cost" (minimum size) collection $\mathscr{C}$ of sets from S such that $\mathscr{C}$ covers all the elements in U. That is, $\bigcup_{S_i \in \mathscr{C}} S_i = U$.

4.2.3 Set Cover Based Greedy Algorithm for DOAM Model

In this section, we first prove that under the DOAM model, the RC-D problem is equivalent to the SC problem. Following the seminal result of [64] that the SC problem is NP-hard, we propose an approximation algorithm called *Set Cover Based Greedy* (SCBG) algorithm for the RC-D problem.

In the following, we show the equivalence between the RC-D problem and the SC problem under the DOAM model, and subsequently, we propose the SCBG algorithm for the RC-D problem.

4.2.3.1   Performance for the RC-D Problem

Theorem 4.1 ([64]) *There is a polynomial time $O(\ln n)$-approximation algorithm for the RC-D problem, where n is the number of bridge ends B.*

Proof Assume that we have an input of RC-D instance $\mathscr{A}$. For each vertex v_i of B, use BFS (Breadth First Search) method to find all vertices that can reach v_i before v_i is infected, this can be done in polynomial time. Assume we have a candidate root set S, for each vertex r_j in S, use BFS method to find all vertices of B that are reachable from r_j before they are infected. Obviously, each root can protect a subset of vertices of B, then the problem becomes a SC problem, i.e., use the least number of roots to cover all vertices of B. Therefore, it has a polynomial time $O(\ln n)$-factor approximation, where n is the number of nodes in B.

Theorem 4.2 *If the RC-D problem has an approximation algorithm with ratio $k(n)$ if and only if the SC problem has an approximation algorithm with ratio $k(n)$.*

Proof Assume $S_1, \cdots, S_m$ is the list of sets for the SC problem and $S_1 \bigcup S_2 \bigcup \cdots \bigcup S_m = \{a_1, \cdots, a_n\}$, we construct a social network as follows.

1. For each set S_i, create a vertex u_i. For each a_j, create a vertex v_j, add directed edges from u_i to v_j if $a_j \in S_i$. An edge from u_i to v_j means v_j can be protected by u_i.
2. Create a social network with a constant number of individuals and an infected node r, add directed edges from r to $v_1, v_2, \cdots, v_n$.
3. Let B be the set of bridge ends including vertices $v_1, v_2, \cdots, v_n$ that need to be protected.
4. The SC problem is converted into the RC-D problem. Thus, it is reasonable to point out that the RC-D problem has a $k(n)$-approximation if and only if the SC problem has a $k(n)$-approximation.

Corollary 4.1 *There is no polynomial time $o(\ln n)$-approximation for the RC-D problem unless $P = NP$.*

Proof It follows from Theorem 4.2 and the well-known inapproximability result for the SC problem [64].

4.2.3.2   The SCBG Algorithm

Now we introduce the SCBG algorithm described in Algorithm 2. The main idea is that we first convert the RC-D problem into the SC problem, then, we apply the greedy algorithm used for the SC problem to obtain the initial protectors for our problem.

The brief description is as follows: given the initial rumor set S_r and bridge end set B. For each node $v \in B$, by using BFS method, we construct v's Bridge End Backward Search Tree (BBST) T_v, in which v is the root of the tree. Denote by $T_1, T_2, \cdots, T_{|B|}$ the BBSTs for corresponding bridge ends. Here $1, 2, \cdots, |B|$

represent the roots of these T_vs, respectively. For $u \in T_i \backslash (S_r \bigcup_{k=0}^{i-1} T_k)$, define $T_0 = \emptyset$, search $T_{i+1}, \cdots, T_{|B|}$ to find the ones that contain u, and record all their corresponding roots and i's corresponding root in the set $T R_u$. Finally, we apply Algorithm 1 to select sets from $T R_u$'s to cover all the nodes in B, and for all these selected sets $T R_v$s, all these v's form the solution to the RC-D problem.

To simplify our expression, here we define the ith level out-neighbors of a node u: let $N^0(u) = u$, and $N^i(u) = N(N^{i-1}(u))$. Since we know the first level out-neighbors of a node, we can easily get the ith level out-neighbors of a node.

Algorithm 1 Greedy algorithm in SCBG algorithm

Input: B, T_i and $T R_j$, where $i = 1, \cdots, |B|$, $j = 1, \cdots, |\bigcup_{k=1}^{|B|} T_k \setminus S_r|$
Output: S_p.
   Initialize $L = \emptyset$ and $S_p = \emptyset$
   while $|L| < |B|$ **do**
      Select $u = \arg\max_{v \in \bigcup_{k=1}^{|B|} T_k \setminus S_r} |T R_v \setminus L|$
      $S_p = S_p \cup \{u\}$ and $L = L \cup T R_u$
   end while
   return S_p.

Algorithm 2 SCBG algorithm-select initial protectors

Input: A directed graph $G = (V, E, C)$, a given community C_m and a set of initial rumors $S_r = \{r_1, r_2, \cdots, r_M\} \subseteq V(C_m)$;
Output: Initial protectors $S_p \subseteq V$;
   for all $r \in S_r$ **do**
      construct *Rumor Forward Search Tree* (RFST) by BFS method to find all bridge ends in G, which are the leaves of the RFSTs, and denote them by a set B;
   end for
   for all node $v \in B$ **do**
      construct *Bridge End Backward Search Tree* (BBST) by BFS method to find all the protector candidates,
      record all the in-neighbors $x \in N^i(v)$ of v, where i is determined by the value of the shortest paths between v and any node $y \in S_r$,
      Denote all the nodes in this tree as a set T_v;
   end for
   List all T_vs as $T_1, \cdots, T_{|B|}$.
   for all $u \in T_i \backslash (S_r \bigcup_{k=0}^{i-1} T_k)$ **do**
      define $T_0 = \emptyset$,
      search $T_{i+1}, \cdots, T_{|B|}$ to find the ones that contain u,
      record all their corresponding roots and i's corresponding root in the set $T R_u$;
   end for
   Apply **Algorithm 1** on $T R_u$s to cover B;
   return Output of **Algorithm 1**.

Fig. 4.1 (a) Rumor
community and its
R-neighbor communities, red
nodes are rumors and green
nodes are bridge ends; (**b**)
Initial protectors v1 and R1
for bridge ends in R-neighbor
communities C1 and C2

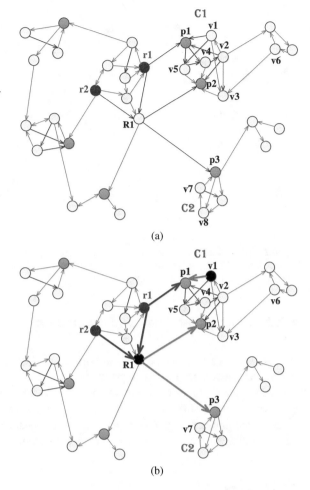

(a)

(b)

We use Fig. 4.1 to show the bridge ends and the corresponding initial protectors
for them. In Fig. 4.1a, the red nodes $r1$ and $r2$ are initial rumors. All green nodes
are bridge ends. In Fig. 4.1b, for simplification, we only illustrate an optimal initial
protectors for R-neighbor communities $C1$ and $C2$, respectively, which are black
vertices $R1$ and $v1$. As seen from Fig. 4.1b, among rumor community and its two
R-neighbor communities $C1$ and $C2$, the green edges form the paths generated by
cascade P ($R1$ and $v1$ are the initial protectors), while the red edges form the paths
generated by cascade R ($r1$ and $r2$ are the initial rumors). Figure 4.2a is Forward
search tree for rumor $r1$ with respect to Fig. 4.1a, and Fig. 4.2b is Backward search
tree for bridge end $p2$ with respect to Fig. 4.1a.

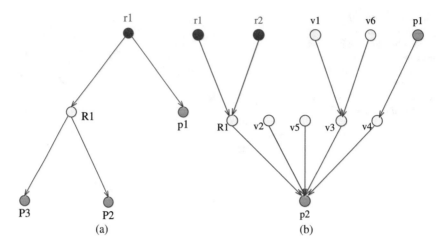

Fig. 4.2 (a) Forward search tree for rumor $r1$ with respect to Fig. 4.1a, and bridge ends are $p1$, $p2$, $p3$; (b) Backward search tree for bridge end $p2$ with respect to Fig. 4.1a, all nodes in this tree except $r1$ and $r2$ can protect $p2$

4.2.4   Experiment Setup and Evaluation

We execute experiments on our algorithms as well as two heuristics in two real-world networks. Our experiments aim at valuating our algorithms from the following aspects: (a) effectiveness with respect to different network density, where network density means the average node degree; (b) effectiveness with respect to different community size, where community size denotes the number of nodes in this community; (c) effectiveness with respect to different number of initial rumors.

4.2.4.1   Datasets

We obtain data from two real-world networks. One network, namely Enron Email communication network, is the same as used in [103, 116]. The other is a collaboration network, which is used in the experimental study in [114], and this network has been shown to capture many key features of social networks in [143].

Enron Email Communication Network

This network covers all the email communications within a dataset of around half million emails. Nodes of the graph represent email addresses and a directed edge from i to j means i sends at least one email to j. This dataset contains 36,692 nodes connected by 367,662 edges with an average node degree of 10.0.

Collaboration Network

Hep collaboration network is extracted from the e-print arXiv, and covers scientific collaborations between authors with papers submitted to High Energy Physics. In this network, nodes stand for authors and an undirected edge between i and j implies that i co-authors a paper with j. Since our problems are based on directed graph, we represent each undirected edge (i, j) by two directed edges (i, j) and (j, i). This dataset contains 15,233 nodes connected by 58,891 edges with an average node degree of 7.73.

To run our experiments, first, we need to obtain the community structure of a social network, since the community partition problem is not a main point in our work, we use a community partition approach proposed by Blondel et al. in [21], and the performance of this approach has been verified in [110]. After obtaining the community structure of a network, we choose different sizes of rumor communities and compute the number of corresponding bridge ends from the two networks. From the Enron Email network, we select two communities, one with 2631 nodes and 2250 bridge ends, and the other with 80 nodes and 135 bridge ends. From the collaboration network, we select a community with 308 nodes and 387 bridge ends.

Finally, we evaluate the performance of our algorithms in comparison with two heuristics: MaxDegree and Proximity. The experimental results are shown in two aspects: (1) Number of selected protectors under the DOAM model; (2) Number of infected nodes under the DOAM model.

We compared the following algorithms to confirm the effectiveness of our algorithms.

MaxDegree A basic algorithm, which simply chooses the nodes according to the decreasing order of node degree as the initial protectors.

Proximity A simple heuristic algorithm, in which the direct out-neighbors of rumors are chosen as the initial protectors.

We do not include the random algorithm due to its poor performance. Instead, a NoBlocking line is included to reflect the performances of these algorithms.

4.2.4.2 Experimental Results

To simplify our presentation, we denote by $|R|$ the number of the initial rumors, $|P|$ the number of the initial protectors, $|C|$ the number of nodes in the rumor community, $|B|$ the number of the bridge ends, $|N|$ the number of nodes in the entire network. To show the simulation results clearly, we adopt the log-time chart.

Table 4.1 Comparison results under the DOAM model

Dataset/$|N|$/$|C|$	$|R|$	SCBG	Proximity	MaxDegree
Hep/15233/308	1%$|C|$	32.9	25.3	140.6
	5%$|C|$	42.1	74.3	147.8
	10%$|C|$	48.9	133.8	152.6
Email/36692/80	5%$|C|$	6.2	43.7	72.7
	10%$|C|$	8.2	46.9	79.3
	20%$|C|$	13.8	62.9	91.1
Email/36692/2631	1%$|C|$	20.4	289.3	1208.8
	5%$|C|$	50.9	1067.6	1350.2
	10%$|C|$	68.4	1422.6	1683.8

Number of Selected Protectors Under the DOAM Model

In Table 4.1, for each rumor community and fixed number of initial rumors (selected randomly), each decimal represents the average number of initial protectors selected by each algorithm (we randomly choose initial rumors for several times and each time we can get a solution). You can see that our SCBG algorithm almost selects the least number of initial protectors no matter where the community is selected and how many initial rumors in it. There is only one exception, in which the rumor community is selected from the Hep network, and has 308 nodes with 3 initial rumors. The reason is that the average node degree is low in Hep network. When the number of initial rumors is pretty small, only a few initial protectors are needed to control the spread of these rumors. Therefore, choosing the direct neighbors of initial rumors is an efficient strategy, that is, Proximity is a good choice.

Furthermore, we also notice that Proximity always performs better than MaxDegree, it is because that Proximity pay attention to the location of initial rumors, thus it can control rumor propagation before they infect a large number of nodes; while MaxDegree only focuses on current influential nodes (nodes having high degree) regardless of the initial rumors. Therefore, it has to choose more initial protectors than Proximity under regular situations. Meanwhile, we also observe that the performance difference among these three algorithms varies under different situations.

Note that among these three communities, the number of initial protectors selected by our algorithms varies much less than that in the other two heuristics. Particularly, in the third community, which is selected from the email network, and has 2631 nodes, when the number of initial rumors increases from 27 (1%$|C|$) to 132, the number of initial protectors selected by our algorithm increases from 20.4 to 50.9 (average value), with the absolute change of 30.5. However, the change in the number of initial protectors is 778.3 and 141.4 for Proximity and MaxDegree, respectively. The results in this community clearly shows that the SCBG algorithm significantly outperforms both Proximity and MaxDegree in networks with large number of nodes and high average node degree.

Number of Infected Nodes Under the DOAM Model

In this part, we focus on testing the effectiveness of these algorithms in protecting nodes in the entire social networks. In other words, for the same number of initial protectors, we want to evaluate the performance of these algorithms. To do this, firstly, for different test cases (different community sizes with different initial rumor sizes), we determine the numbers of initial protectors, respectively, and these numbers are slightly larger than those selected by the SCBG algorithm. Then, for each test, from corresponding solutions, we randomly choose predetermined size of nodes as initial protectors. Thirdly, we run the three algorithms using selected initial protectors. Since each predetermined number is larger than the number of nodes selected by our algorithm, besides using the nodes in its solution, our algorithm also has to use some randomly selected nodes. From Figs. 4.3, 4.4, and 4.5, we observe that rumors propagate very fast within the first four steps while after the fourth step, almost no new nodes are infected over all test cases.

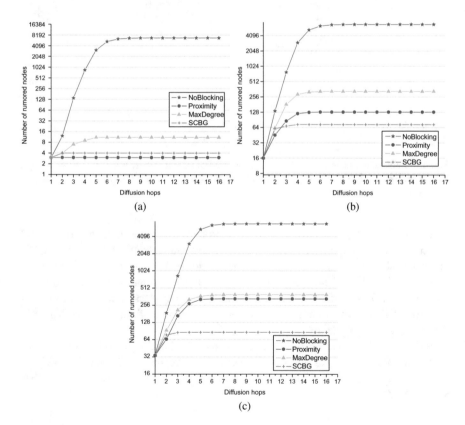

Fig. 4.3 Infected nodes under the DOAM model on Hep collaboration network with $|N| = 15,233$, $|C| = 308$, $|B| = 387$. (a) $|R| = 1\%|C|$, $|P| = 34$. (b) $|R| = 5\%|C|$, $|P| = 44$. (c) $|R| = 10\%|C|$, $|P| = 55$

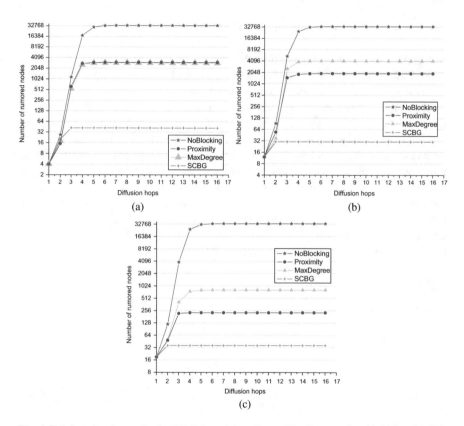

Fig. 4.4 Infected nodes under the DOAM model on Enron Email network with $|N| = 36,692$, $|C| = 80$, $|B| = 135$. (**a**) $|R| = 5\%|C|, |P| = 8$. (**b**) $|R| = 10\%|C|, |P| = 11$. (**c**) $|R| = 20\%C, |P| = 14$

Except Fig. 4.3a, in which the Proximity protects one more node than the SCBG algorithm due to small size of initial rumors and low network density, the SCBG algorithm always protects the most number of nodes in comparison with the other two heuristics. Therefore, we believe that our algorithm can be applied to those problems that aim at either protecting targeted nodes with least number of protectors or reducing the number of nodes infected in the entire networks at the end of cascade diffusion, or both of them.

We also notice that Proximity outperforms MaxDegree for different sizes of initial rumors in Figs. 4.3 and 4.4. However, we can see in Fig. 4.5, MaxDegree performs better than Proximity. The reason is that this network has much higher average node degree.

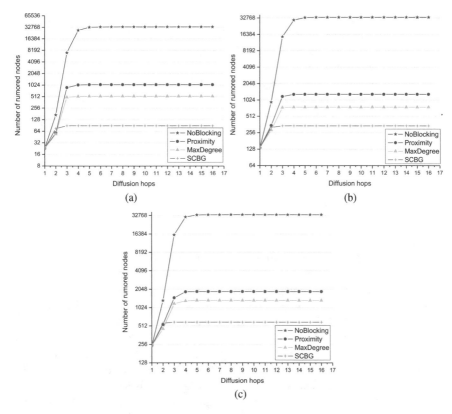

Fig. 4.5 Infected nodes under the DOAM model on Enron Email network with $|N| = 36,692$, $|C| = 2631$, $|B| = 2250$. (a) $|R| = 1\%|C|$, $|P| = 21$. (b) $|R| = 5\%|C|$, $|P| = 52$. (c) $|R| = 10\%|C|$, $|P| = 69$

4.3   Rumor Blocking Maximization with Constrained Time

In this section, we seek effective strategy to stop the diffusion of rumor in a network considering the following three factors:

1. A constraint on how much time we can use to control the spread of rumor in the network.
2. There is usually a random time delay in influencing a friend when a person accepts new information.
3. Individuals make decisions based on relationships with their informed friends, but also on their personal judgement about the piece of information that is being diffused.

Here we propose two general models to capture diffusion of rumor and truth (protector) in a social network and formally define the rumor containment problem under these two models as two optimization problems. NP-hardness results are established for these two optimization problems. We prove the submodularity of the objective functions in these two problems. This enables us to use the greedy algorithm as a constant-factor approximation algorithm (for both problems) with a performance guarantee $1 - \frac{1}{e}$.

The rest of this chapter is organized as follows: Sect. 4.3.1 presents the two propagation models. In Sect. 4.3.2, we define the rumor blocking maximization problem formally, prove its NP-hardness under the two models, and establish the submodularity of the objective functions. Section 4.3.3 gives the formal description of the greedy algorithm.

4.3.1 Propagation Models

Social network can be modeled as a directed graph $G = (V, E)$, where V is the node set and E is the edge set. In the context of influence diffusion, V represents the individuals in this network and E represents the relationships among these individuals. Furthermore, a node $v \in V$ is an out-neighbor of a node $u \in V$ if there exists an edge $e_{uv} \in E$ (i.e., the edge from node u to node v exists in graph G). A node u is called an in-neighbor of v if v is an out-neighbor of u.

In reality, two individuals in a network may not interact/exchange information every day. If someone gets influenced by a certain event, then her friends may learn about this fact *several* days later, and get influenced also. In other words, there is a random time delay between the influence friends have on each other. In this paper, we model this phenomenon using the meeting probability among two nodes in the graph: the meeting action between a node u and its neighbor v happens stochastically at any time step with probability m_{uv}, independent of everything else. Moreover, each edge e_{uv} is assigned an influence weight (probability) IW_{uv}. In the following two models, we will explain this influence weight separately in detail.

4.3.1.1 Rumor-Protector Independent Cascade Model with Meeting Events (RPIC-M)

We first describe one of the most basic and well-studied diffusion models in [93], namely the independent cascade (IC) model. Then, we describe a generalized model which models the following additional features: competitive influence diffusion, meeting events, and personal interest.

In the IC model, a network is considered as a directed graph $G = (V, E)$, where V denotes individuals in the network and E denotes the relationship between individuals. Each edge $e_{uv} \in E$ is assigned an influence probability p_{uv}, indicating the possibility that node u influences node v successfully. For $e_{uv} \notin E$, let $p_{uv} = 0$. Each node can only be in one the following two statuses: inactive or active. Once a node becomes active, it will remain active forever. The diffusion process unfolds in discrete time steps. Starting with an initial set of active nodes A_0, at any step $t \geq 1$, when node u first becomes active in step t, it has a single chance to activate any of its currently inactive neighbors. For neighbor node v, it succeeds with probability p_{uv}. If u succeeds in activating v, then v will become active in steps $t + 1$, and if u fails in activating v, then v will remain inactive. Regardless of the activation outcome, u cannot make any further attempts to activate v in all subsequent rounds. The process continues until no more activations are possible. If multiple newly activated nodes are in-neighbors of the same inactive node, then their attempts are sequenced in an arbitrary order.

We now describe our new model that incorporates competitive influence (the influence of protector and rumor) diffusion, as well as meeting events and personal interest. This model extends the models proposed in [30]. We denote it by RPIC-M (Rumor-Protector Independent Cascade model with Meeting events). Let P (for "protector") and R (for "rumor") denote the two cascades. The initial set of protected (resp., infected) nodes is denoted by A_p and (resp., A_r). Each node u has personal interests in the information (PI_u). This parameter PI_u is a probability and plays an role in activating node u. Each node is either inactive, infected, or protected. Each edge e_{uv} is associated with a meeting probability m_{uv} and an influence probability p_{uv} (if $e_{uv} \notin E, m_{uv} = 0$ and $p_{uv} = 0$).

Given rumor seed set A_r, as in the IC model, a protector seed set A_p is selected and activated at step $t = 0$. At any step $t \geq 1$, a protected (resp., infected) node u meets any of its currently inactive neighbors v independently with probability $m(u, v)$. Since u's activation, if a meeting event happens between u and v for the first time, then u has a single chance to try protecting (resp., infecting) v with an influence probability $\min\{1, p_{uv} + PI_v\}$, given that no other neighbor of v tries protecting or infecting v at the same step.

If the attempt from u succeeds, v becomes influenced (protected or infected) at step t and will start influencing (protect or infect) its inactive neighbors from time $t + 1$ onwards. If there are two or more nodes trying to influence v simultaneously, at most one of them can succeed. The attempts from the same cascade are ordered arbitrarily. As for the attempts from different cascades (P and R), we assume that all the attempts in R have priority over P. Once a node becomes protected or infected, it will never change its status. The diffusion process continues until no more nodes can be protected or infected.

4.3.1.2 Rumor-Protector Linear Threshold Model with Meeting Events (RPLT-M)

Again, we first describe another basic and well-studied diffusion models in [93], namely the linear threshold (LT) model. Next, we describe a generalized model which models the following additional features: competitive influence diffusion, meeting events, and personal interest.

In the LT model, a social network is viewed as a directed graph $G = (V, E)$, where V denotes individuals in the network and E denotes the relationship between individuals. Each edge $e_{uv} \in E$ is assigned a non-negative weight w_{uv}, which represents the impact that node u has on node v. For $e_{uv} \notin E$, let $w_{uv} = 0$. For each $v \in V$, $\sum_{u \in V} w_{uv} \leq 1$. Each node v is associated with a random threshold θ_v, which is drawn independently from a uniform distribution with support $[0, 1]$. Each node is either active or inactive. Once a node becomes active, it remains active forever. The diffusion process unfolds in discrete time steps. Starting with an initial set of active nodes A_0, at any step $t \geq 1$, a node v will become active if and only if the total weight coming from its active in-neighbors exceeds its threshold θ_v, i.e., $\sum_{u \in A_{t-1}} w_{uv} \geq \theta_v$, where A_{t-1} is the set of active nodes by time step $t - 1$. This diffusion process continues until no more nodes can be activated.

We now describe our new model that incorporates competitive influence (the influence of protector and rumor) diffusion, as well as meeting events and personal interest. This model extends the models proposed in [84]. We denote it by RPLT-M (Rumor-Protector Linear Threshold model with Meeting events). Let P (for "protector") and R (for "rumor") denote the two cascades. The initial set of protected (resp., infected) nodes is denoted by A_p and (resp., A_r). Each node v has personal interests in the information from protector (PI_{pv}) and rumor (PI_{rv}). These two parameters are probabilities and play a role in activating node v. Each node is either inactive, infected, or protected. Each edge e_{uv} is associated with a meeting probabilities m_{uv} as well as two weights $w_{uv,p}$ and $w_{uv,r}$ (if $e_{uv} \notin E$, $m_{uv} = 0$ and $w_{uv,p} = w_{uv,r} = 0$). We assume that for all node $v \in V$, $\sum_{u \in V} w_{uv,p} + PI_{pv} \leq 1$ and $\sum_{u \in V} w_{uv,r} + PI_{rv} \leq 1$. Each node u chooses two independent thresholds, namely θ_{pu} (for P) and θ_{ru} (for R) randomly from the uniform distribution with support $[0, 1]$.

Given rumor seed set A_r, as in the LT model, a protector seed set A_p is selected and activated at step $t = 0$. At any step $t \geq 1$, a protected (resp., infected) node u keeps its status and meets any of its currently inactive neighbors v with probability $m(u, v)$. Since u's activation, if a meeting event happens between u and v for the first time, then we say that u's influence (protect or infect) to v is valid. An inactive node v is protected (resp., infected) if the total valid weight from its protected (resp., infected) in-neighbors plus its own interest PI_{pv} (resp., PI_{rv}) exceeds its threshold θ_{pv} (resp., θ_{rv}), given that v has not been activated (infected or protected) yet. If at step t, v is both successfully influenced by P and R, then the diffusion of R has priority over that of P, and v becomes protected. Once a node becomes protected or infected, it will never change its status. The diffusion process continues until no more nodes can be protected or infected.

4.3.2 Rumor Containment with Constraints

In Sect. 4.3.2.1, we define the problem of Rumor Containment maximization with the following additional constraints: time Deadline, Meeting events, and Personal interests (RC-DMP). Subsequently in Sect. 4.3.2.1, we show that RC-DMP under the RPIC-M and RPLT-M models are both NP-hard. Finally in Sect. 4.3.2.1, we prove that the objective functions of RCM-DM under the above-mentioned two models are monotone and submodular. Following the seminal result of [141], the greedy algorithm is a constant-ratio approximation algorithm for RC-DMP with performance guarantee $1 - \frac{1}{e}$.

4.3.2.1   Problem Definition

We note that previous research on rumor blocking (see, e.g., [30, 62, 84, 147]) fails to address the following additional features:

1. there is often a time deadline on how long the diffusion process can last.
2. people do not interact with each other (influence each other) every day (i.e., the influence between individuals happens randomly, instead of deterministically at every time step).
3. In addition to the influence coming from an individual's friends, she has her own personal interests/opinions about the information that is being diffused. This may also affect how well she accepts the information.

We consider these three factors in our RC-DMP problem.

The problem is formally defined as follows: given a directed graph $G = (V, E)$, a rumor seed set A_r, and two positive integers k and T, our goal is to find a protector seed set A_p ($|A_p| \leq k$) to minimize the expected number of infected nodes by the end of time deadline T. We denote the objective function for the RC-DMP problem by $RC_T(S)$, which is the number of nodes that will be infected within deadline T by the diffusion of rumor, if instead of the set S, the empty set is chosen as the protector seed set.

NP-Hardness of RC-DM

In this section, we prove that the RC-DMP problem under our two proposed models is NP-hard.

NP-Hardness of RC-DMP Under the RPIC-M Model

Theorem 4.3 *Problem RC-DMP under the RPIC-M model is NP-hard.*

Proof Consider the following special case of Problem RC-DMP: The time deadline $T = +\infty$, all the meeting probabilities are equal to 1, all the personal interests are 0 and $p_{uv} = 1$ for all $e_{uv} \in E$. We note that this special case of Problem RC-DMP is identical to the Problem LCRB-D considered in [62] except the tie-breaking rule. A similar reduction from the Set Cover problem as that in the proof of Theorem 3 of [62] can be used to prove this result.

Next, we describe the reduction formally. Given an integer k, a ground set $N = \{m_1, m_2, \ldots, m_n\}$ and a list of subsets of N: $S_1, S_2, \ldots, S_m$ such that $S_1 \cup S_2 \cup \ldots S_m = \{m_1, m_2, \ldots, m_n\}$, the Set Cover problem wishes to find k subsets of N from the list, such that the union of these subsets covers the entire ground set. We reduce the Set Cover problem to our problem by constructing a directed graph as follows:

1. For each subset S_i, $i = 1, 2, \ldots, m$, create a vertex u_i. For each element m_j, $j = 1, 2, \ldots, n$, create a vertex v_j. Add a directed edge from u_i to v_j if $m_j \in S_i$.
2. Create a rumor node r. Add directed edges from r to $u_1, u_2, \ldots, u_m$.

Now it is easy to see that the set cover instance is a "yes" instance if and only if in our problem we can find a S such that $RC_T(S) \geq n + k$. The result follows.

NP-Hardness of RC-DMP Under the RPLT-M Model

Theorem 4.4 *Problem RC-DMP under the RPLT-M model is NP-hard.*

Proof Consider the following special case of Problem RC-DMP: The time deadline $T = +\infty$, all the meeting probabilities are equal to 1, and all the personal interests are 0. We note that this special case of Problem RC-DMP is identical to the Problem IBM under the CLT model considered in [84], which is shown to be NP-hard. The result follows.

Submodularity of RC-DMP A set-based function $f : 2^S \rightarrow \mathbb{R}$ is called submodular if it has the property of diminishing marginal returns, that is, $f(A \cup \{u\}) - f(A) \geq f(B \cup \{u\}) - f(B)$, $\forall A \subseteq B \subset S$, $\forall u \in S \setminus B$. Furthermore, f is monotone if it satisfies $f(A) \leq f(B)$ when $A \subseteq B \subset S$. In the following, we prove that the objective functions of our RC-DMP problem under the RPIC-M and the RPLT-M models are monotone and submodular. To maximize a non-negative, monotone, and submodular function, we can use the well-known greedy hill-climbing algorithm [141] to obtain a constant approximation ratio of $1 - \frac{1}{e}$.

Submodularity of RC-DMP Under the RPIC-M Model

Theorem 4.5 *Function $RC_T(\cdot)$ is monotone and submodular for any instance of RC-DMP under the RPIC-M model.*

Proof Similarly to the proof in [93], we establish the "*live-path*" graph to demonstrate the submodularity of our objective function. Since the cascade process under the RPIC-M model is random, we can suppose that before the cascade starts, a set of outcomes for all meeting events as well as the *live* or *blocked* assignment for all edges are already determined. The "live-path" graph G_{live} is constructed by combining the two outcomes. Specifically, a live edge e_{uv} is added to G_{live} in the event that u is activated (infected or protected) and is meeting the inactive v for the first time.

For each meeting event (an edge e_{uv} and a time step t in $[1, T]$), we flip a coin with bias m_{uv} to determine if u will meet v at t. Similarly, for each edge e_{uv}, we flip a coin once with bias $\min\{1, p_{uv} + PI_v\}$, and we declare the edge "live" with probability $\min\{1, p_{uv} + PI_v\}$, or "blocked" with probability $1 - \min\{1, p_{uv} + PI_v\}$. All the two operations with coin-flips are independent.

Given an instance S_M of outcomes of all meeting events ($\forall e_{uv} \in E, \forall t \in [1, T]$), and also an instance S_{LB} of live or blocked assignments for all edges, since the process for meeting events and that of the live or blocked assignment are different, and moreover, all flips in the two processes are independent, a possible instance S of all the random outcomes of our problem can be obtained by combing S_M and S_{LB}.

For a fixed S, the two cascades unfold deterministically. Let $DM_T^S(A)$ denote by the end of time step T, the node set that will be infected if instead of A, the empty set is chosen as the initial protector seed set. Note that by definition, we have that

$$RC_T(A) = \sum_S Prob(S) \cdot |DM_T^S(A)|.$$

In the classic IC model, given outcome S, for a live edge e_{uv} in the graph, node u can reach node v with one hop. However, in our model with meeting events, given outcome S, for a live edge e_{uv} in the graph, u will reach node v with $t_v - t_u$ hops, where t_u is the step in which u itself is activated, and t_v is the first step when u meets v, after t_u.

Hence, we say that v is reachable from a seed set A if and only if

- There exists at least one path consisting entirely of live edges (called live-path) from some node in A to v.
- The collective number of hops along the shortest live-path from S to v is no greater than T.

For any given outcome S, consider the graph $G_{live} = (V, E')$, where V is the vertex set of graph G, and E' is the set of live edges in E (determined by S). Both rumor and protector can propagate in this graph. Let V' denote the nodes that can be reached by rumor seed set A_r via live edges within T time steps. Then we construct another graph $G' = (V'', E'')$, where $V'' = \{v | v \in V \text{ and } v \notin A_r\}$ and $E'' = \{e_{uv} | u, v \in V'' \text{ and } e_{uv} \in E'\}$. Since the rumor seed set is given and the meeting event at each time step for pair of nodes is determined by S, we can determine the time step t_u that $u \in V'$ is infected. Similarly, for a protector seed set A_p, we can also determine the time step t_u' that $u \in V'$ is protected.

To construct the protector reachability graph, we do as follows: If $t_u' < t_u$, then we keep the live-path from A_p to u. Otherwise, we delete the path. For all the nodes in V', we determine whether there exists a live-path from A_p. Let $A \subseteq B \subseteq V''$, consider the quantity $|DM_T^S(A \cup \{u\})| - |DM_T^S(A)|$. This is the number of nodes that can be reached by node u but cannot be reached by any node in set A. This is at least as large as the number of nodes that can be reached by node u but cannot be reached by any node in set B. In other words, $|DM_T^S(A \cup \{u\})| - |DM_T^S(A)| \geq |DM_T^S(B \cup \{u\})| - |DM_T^S(B)|$, indicating that $|DM_T^S(\cdot)|$ is submodular. Taking expectation over all possible S, we conclude that the function $RC_T(\cdot)$ is also submodular.

Submodularity of RC-DMP Under the RPLT-M Model

We follow the general idea in [93] for the proof, that is, we prove that the influence diffusion process guided by the RPLT-M model is equivalent to the one guided by a random live or blocked assignment process. Since we have meeting events in our model, we need to incorporate them into the live or blocked assignment process. We now describe the live or blocked assignment process that we use.

Since the meeting event associated with each edge is random, we can determine them for each edge e_{uv} at any time step t by pre-flipping a coin. Given an outcome

S_M of all the random meeting events, for each edge $e_{uv} \in E$, the outcome of meeting events at each time step is determined by S_M. Based on the original graph $G = (V, E)$, we construct two random graphs, namely $G_{S_M R} = (V, E_R)$ and $G_{S_M P} = (V, E_P)$ for rumor diffusion and protector diffusion, respectively.

To construct $G_{S_M P}$, for each node $v \in V$, with probability $w_{uv,p}$, only in-edge e_{uv} is selected and marked as live. With probability PI_{pv}, we mark all of its in-edges as live, and with probability $1 - (\sum_{u \in V} w_{uv,p} + PI_{pv})$ no in-edge is selected as live. (Note that our live-or-blocked assignment process differs with [93] in the sense that for a node v, multiple in-edges can be selected as live. While in [93], at most one edge can be selected as live.)

Similarly, to obtain graph $G_{S_M R}$, for each node $v \in V$, with probability $w_{uv,r}$, only in-edge e_{uv} is selected and marked as live. With probability PI_{rv}, we mark all of its in-edges as live, and with probability $1 - (\sum_{u \in V} w_{uv,r} + PI_{rv})$ no in-edge is selected as live.

We define the concept of an effective live edge as follows: At any step t, live edge e_{uv} becomes *effective* when v meets with its selected neighbor u for the first time, and u has been activated at some earlier step $t' < t$.

In $G_{S_M R}$ (resp. $G_{S_M P}$), given a rumor seed set A_r (resp. protector seed set A_p), for a node $v \in V$, if its selected live edges for rumor diffusion (resp. protector diffusion) connect some node u in A_r (resp., A_p), and in S_M, u meets v before deadline T, then edge e_{uv} becomes effective. If u is not in A_r (resp., A_p), but u has been influenced at t_{ru} (resp., $t_{pu}^{A_p}$ since A_p is not a fixed set), and in S_M, u meets v before deadline T, then edge e_{uv} also becomes effective. If a node u cannot be activated by rumor diffusion (resp., protector diffusion) by the end of time step T, then we define $t_{ru} = \infty$ (resp., $t_{pu}^{A_p} = \infty$), meaning no effective live rumor path (resp., protector path) exists between A_r and u. We say that a node u is *protected* if $t_{pu}^{A_p}, t_{ru} < \infty$ and $t_{pu}^{A_p} < t_{ru}$, and u is *infected* if $t_{ru} < \infty$ and $t_{ru} \leq t_{pu}^{A_p}$.

The following lemma states that the distribution over the final activated (protected or infected) nodes are identical for our RPLT-M and the above live or blocked assignment process.

Lemma 4.1 *For a given protector seed set A_p and rumor seed set A_r, the distribution over the sets of nodes that are infected and protected is identical in the following two models:*

1. *RPLT-M model.*
2. *the live or blocked assignment process.*

Proof We prove this lemma by proving this equivalence under any fixed outcome S_M of the meeting events.

To proceed, we first look at the diffusion process under the RPLT-M model for a given S_M. Recall that the diffusion unfolds in discrete time steps. In each step, some nodes change from inactive to active (protected or infected). For all $t \in [0, T]$, let $A_t^p(v)$ be the set of nodes that are already protected and have met v at least once after their activation by the end of step t, and $A_t^r(v)$ be the set of nodes that are already infected and have met v at least once since their activation by the end of step t. Consider a node v that has not been activated by the end of time step t. The probability that v becomes protected in $t + 1$ equals to the probability that the incremental weight contributed by $A_t^P(v) \setminus A_{t-1}^P(v)$ pushes it over the threshold θ_{pv} (and the incremental weight contributed by $A_t^r(v) \setminus A_{t-1}^r(v)$ does not push it over the threshold θ_{rv}), given that it is not activated by the end of step t. This probability is:

$$\frac{\sum_{u \in A_t^P(v) \setminus A_{t-1}^P(v)} w_{uv,p} \left[1 - \sum_{u \in A_t^r(v) \setminus A_{t-1}^r(v)} w_{uv,r} \right]}{\left[1 - \left(\sum_{u \in A_{t-1}^P(v)} w_{uv,p} + P I_{pv} \right) \right] \cdot \left[1 - \left(\sum_{u \in A_{t-1}^r(v)} w_{uv,r} + P I_{rv} \right) \right]}.$$

Similarly, the probability that node v becomes infected in $t + 1$ given that v is inactive from step 0 to t is:

$$\frac{\sum_{u \in A_t^r(v) \setminus A_{t-1}^r(v)} w_{uv,r}}{\left[1 - \left(\sum_{u \in A_{t-1}^P(v)} w_{uv,p} + P I_{pv} \right) \right] \cdot \left[1 - \left(\sum_{u \in A_{t-1}^r(v)} w_{uv,r} + P I_{rv} \right) \right]}.$$

Next, we look at the live or blocked assignment process for the same fixed outcome S_M of the meeting events. Let B_0^P and B_0^r denote protector seed set and rumor seed set, respectively. For each $t \in [1, T]$, let B_t^P denote the set that contains any $v \notin B_{t-1}^P \cup B_{t-1}^r$ such that v has one effective live in-edge from some node in B_{t-1}^P but no effective live in-edge from any node in B_{t-1}^r. For each $t \in [1, T]$, let B_t^r denote the set containing any $v \notin B_{t-1}^P \cup B_{t-1}^r$ such that v has one effective live in-edge from some node in B_{t-1}^r.

According to the definition of random live or blocked assignment process, the probability that a node v is in $B_{t+1}^P \setminus B_t^P$ conditioned on that v is not in $B_t^P \cup B_t^r$ is:

$$\frac{\sum_{u \in A_t^P(v) \setminus A_{t-1}^P(v)} w_{uv,p} \left[1 - \sum_{u \in A_t^r(v) \setminus A_{t-1}^r(v)} w_{uv,r} \right]}{\left[1 - \left(\sum_{u \in A_{t-1}^P(v)} w_{uv,p} + P I_{pv} \right) \right] \cdot \left[1 - \left(\sum_{u \in A_{t-1}^r(v)} w_{uv,r} + P I_{rv} \right) \right]}.$$

Similarly, the probability that a node v is in $B_{t+1}^r \setminus B_t^r$ conditioned on that v is not in $B_t^P \cup B_t^r$ is:

$$\frac{\sum_{u \in A_t^r(v) \setminus A_{t-1}^r(v)} w_{uv,r}}{\left[1 - \left(\sum_{u \in A_{t-1}^P(v)} w_{uv,p} + P I_{pv} \right) \right] \cdot \left[1 - \left(\sum_{u \in A_{t-1}^r(v)} w_{uv,r} + P I_{rv} \right) \right]}.$$

The above conditional probabilities are the same as that obtained from the RPLT-M model. Since $A_p = B_0^P$ and $A_r = B_0^r$, we conclude our proof.

With the help of the equivalence result in Lemma 4.1, we can now prove the monotonicity and submodularity of rumor blocking in the random live or blocked assignment process. Given a fixed outcome S_M of the meeting events, a rumor seed set A_r and an instance S_L of the live or blocked assignment process (where the outcomes of all the live edge selections are determined), let $X = (S_M, S_L)$ and $RC_T^X(A)$ denote the node set that will be infected if instead of A, the empty set is chosen as the initial protector seed set. Then the objective function for our rumor blocking problem is

$$RC_T(A) = \mathbb{E}_X |RC_T^X(A)|.$$

Given a graph $G(V, E)$ and a set $S \subset V$, for a node $u \notin S$, we say that there is a unique path from S to u if there exists some path from a node in S to u. For any two paths from any two nodes in S to u, one path must be a sub-path of the other.

Next, we establish the following lemmas to prove the submodularity of $RC_T^X(A)$.

Lemma 4.2 *In an effective rumor path graph $G_{S_M}R$ ($G_{S_M}P$), given a protector seed set A, for any node u, if $t_{ru} < \infty$ (resp., $t_{pu}^A < \infty$), then there is a unique effective rumor (resp., protector) path from some node in A_r (resp., A) to v.*

Lemma 4.3 *The sufficient and necessary condition for $v \in RC_T^X(A)$ is:*

- *There exists a unique effective rumor path from A_r to v;*
- *there exists at least one node u in the unique rumor path, such that a unique effective protector path exists between A and u with $t_{pu}^A < t_{ru}$.*

Lemma 4.4 *The sufficient and necessary condition for $v \in RC_T^X(B \cup \{u\}) \setminus RC_T^X(B)$ is:*

- *There exists a unique effective rumor path from A_r to v;*
- *There exists at least one node w on the unique effective rumor path from A_r to v, such that a unique effective protector path exists between $B \cup \{u\}$ and w with $t_{pw}^{B \cup \{u\}} < t_{rw}$;*
- *for all node x on the unique effective rumor path from A_r to v, it holds that $t_{rx} \leq t_{px}^B$.*

Lemma 4.5 *The cardinality set function $|RC_T^X(A)|$ for an instance $X = (S_M, S_L)$ is monotone and submodular.*

Proof First we show that $|RC_T^X(A)|$ is monotone. That is, for any node $u \in V \setminus (A \cup A_r)$ where $A \subseteq V$, we need to prove that $|RC_T^X(A)| \leq |RC_T^X(A \cup \{u\})|$, which is equivalent to showing that $RC_T^X(A) \subseteq RC_T^X(A \cup \{u\})$. Consider any node $v \in RC_T^X(A)$, we have that $t_{rv} < \infty$, meaning that there exists a node w in the unique effective rumor path from A_r to v such that $t_{pw}^A < t_{rw}$. We also know that $t_{pw}^{A \cup \{u\}} \leq t_{pw}^A$, therefore, we have that $t_{pw}^{A \cup \{u\}} < t_{rw}$. Thus, we have that $v \in RC_T^X(A \cup \{u\})$.

To prove the submodularity of $|RC_T^X(A)|$, we show that for any $A \subseteq B \subseteq V$, and $u \in V \setminus B$, we have that $RC_T^X(B \cup \{u\}) \setminus RC_T^X(B) \subseteq RC_T^X(A \cup \{u\}) \setminus RC_T^X(A)$. That is, we only need to show that for any $v \in RC_T^X(B \cup \{u\}) \setminus RC_T^X(B)$, we have $v \in RC_T^X(A \cup \{u\}) \setminus RC_T^X(A)$. Since we know that there exists a node w

on the unique effective rumor path from A_r to v, and $t_{pw}^{B\cup\{u\}} < t_{rw}$. And for all node x on the effective rumor path from A_r to v, $t_{px}^B \geq t_{rx}$. Therefore, for node w, $t_{pw}^{B\cup\{u\}} < t_{rw} \leq t_{pw}^B$, meaning that the influence from node u can reach w earlier than all the nodes in B and A_r. Therefore, the influence from u can reach w earlier than all the nodes in A, that is, $t_{pw}^{A\cup\{u\}} = t_{pw}^{B\cup\{u\}} < t_{rw}$. Since for any node x on the unique effective rumor path from A_r to v, we have that $t_{px}^B \geq t_{rx}$, and $A \subseteq B$, it is clear that $t_{px}^A \geq t_{px}^B$, thus, $t_{px}^A \geq t_{rx}$, thus, we have demonstrated that $RC_T^X(B \cup \{u\}) \setminus RC_T^X(B) \subseteq RC_T^X(A \cup \{u\}) \setminus RC_T^X(A)$.

Since taking expectation preserves submodularity, we have established the following result.

Theorem 4.6 *Function $RC_T(\cdot)$ is monotone and submodular for any instance of RC-DMP under the RPLT-M model.*

4.3.3 Possible Solutions

From Theorems 4.3 and 4.4, we know that Problem RC-DMP is NP-hard under the two proposed models (RPIC-M and RPLT-M). This motivates our consideration for approximation algorithm for Problem RC-DMP. Moreover, from Theorem 4.6, we know that the objective function $RC_T(\cdot)$ of Problem RC-DMP under the RPIC-M and the RPLT-M models is monotone and submodular. Furthermore, by definition, $RC_T(\cdot)$ is non-negative and $RC_T(\emptyset) = 0$. Consequently, we can apply the seminal result in [141] and use the greedy algorithm as a constant-factor approximation algorithm with performance guarantee ratio of $1 - \frac{1}{e}$. We formally present the greedy algorithm in Algorithm 3. Note that variable R in the algorithm controls the number of Monte Carlo simulations.

Algorithm 3 Greedy algorithm

Input: Given a graph $G = (V, E)$, A_r, k and T
Output: Protector seed set $A_p \subseteq V$.
1: Initialize $A_p = \emptyset$, $R = Num\_Simulations$
2: **for** $i = 1$ **to** k **do**
3:
4:     **for all** $u \in PV \setminus A_p$ **do**
5:       $IF(u) = 0$
6:     **end for**
7:     **for** $j = 1$ **to** R **do**
8:       $IF(u) + = RC_T(A_p \cup \{u\})$
9:     **end for**
10:    $IF(u) = IF(u)/R$
11:    $A_p = A_p \cup \arg\max_{u \in V \setminus A_p}\{IF(u)\}$
12: **end for**
13: Output A_p.

Since in our problem, rumor and protector diffuse with time deadline T, meaning that we only need to search certain area for computation of the seed set of protectors. Let $N_{in}(u)$ denote one hop in-edge neighbors of node u, and $N_{in}^2(u) = \{w | e_{wv} \in E \cap v \in N_{in}(u)\}$ denote two hops in-edge neighbors of node u, thus $N_{in}^t(u)$ are the t hops in-edge neighbors of node u. Moreover, we denote $R^t(A_r) = \bigcup_{u \in A_r} N_{in}^t(u)$ and $R = \bigcup_{t=1}^{t=T} R^t(A_r)$. $P^t = \bigcup_{u \in R^t(A_r)} N_{in}^{t-1}(u)$, where $t \in [1, T]$, and $P = \bigcup_{t=1}^{t=T} P^t$. As a result, $PV = P \cup R$ is the valid nodes that we only need to compute in our objective function.

4.4 Conclusion

In this chapter we performed an extensive study of the problem of limiting the spread of misinformation/rumors in a social network. We investigated efficient solutions to the following question: Given a social network where a (bad) information campaign is spreading, who are the influential people to start a counter-campaign if our goal is to block the effect of the bad campaign efficiently?

In Sect. 4.2, we formulated the Rumor Control (RC) problem under the DOAM model and prove that it is equivalent to the Set Cover problem. To address the problem, Set Cover Based Greedy (SCBG) algorithm is presented, which contains two parts: first, transfer the RC-D problem into the SC problem; second, apply the greedy algorithm used for the SC problem to the obtained subsets for bridge ends. The experimental reports over two real-world social networks demonstrate that the SCBG algorithm outperforms the two heuristics: MaxDegree and Proximity.

In Sect. 4.3, we proposed two models to capture competitive influence diffusion process, namely the RPIC-M and RPLT-M models. In these two models, two kinds of cascades propagate: protector and rumor. These two models extends the seminal IC and LT models [93] to the case of two-cascade influence diffusion. Furthermore, the following three features are also included in these models: a time deadline, random time delay between information exchange, and personal interests regarding the acceptance of information.

Under these two models, we study the RC-DMP problem: given a directed graph $G = (V, E)$, a rumor seed set A_r, and two positive integers k and T, our aim is to find a protector seed set A_p (with $|A_p| \leq k$) to minimize the expected number of infected nodes by the end of time deadline T. We prove that the problem under the two models is both NP-hard. Moreover, we demonstrate that the objective functions under the two different models are both monotone and submodular. Therefore, we are able to apply the seminal result in [141] and use the greedy algorithm as a constant-factor approximation algorithm with performance guarantee ratio of $1 - \frac{1}{e}$.

About future directions, we mention several clues. First, the greedy approximation algorithm is inefficient and time-consuming as it lacks of a way to efficiently compute the objective functions for our problem. To overcome such inefficiency, we hope to find more efficient algorithms to compute the objective function under the two proposed models.

Second, we have noticed that under most situations, the spread of influence and the meeting events occurred among individuals are in continuous time. Thus developing continuous-time diffusion models for our problem is promising.

Third, more real-world factors such as personal interests, different influence diffusion speed, deadline, etc., could be incorporated into current diffusion models.

Last but not the least, in society, influence diffuses in different mechanisms, as well as in different contexts, that is, there exist various models in reality. Therefore, it is interesting to look into our problem under other influence diffusion models.

Chapter 5
Multiple Social Influence: Models and Applications

5.1 Overview

Cascading processes are models of network diffusion used to study phenomenon concerning the spread of new trends and innovations in social networks. Each node can be in one of two states: infected (i.e., supports an idea or a product) or uninfected. Every infected node can infect its neighbors and thus, the infection, formally called a cascade, propagates through the network. These processes have been studied in many applications such as viral marketing [53], blog networks [114] and contagion models [51].

Broadly two theoretical models of diffusion have been explored: the linear threshold model [78] and the independent cascade model [73] (please refer to Chap. 2). In the former, every infected neighbor for a node contributes certain weights and if their sum is greater than a threshold, the node is infected. The weights depend often on the edge strength between the node and its neighbors. In the latter, each infected node is allowed one chance to infect a neighbor with some probability generally depending on the edge strength between the nodes.

Existing literature has primarily focused on single cascade models but this assumption breaks down in many real-world scenarios when there are many competing products, different political messages, ideas, etc. It is also possible for nodes' affinities towards certain cascades to evolve with those of their neighbors. This situation has different dynamics and requires more sophisticated models. Research in two competitive cascades has looked at variations of the independent cascade model (please refer to previous chapter).

In this chapter, we discuss models and applications where multiple social cascades (n cascades) propagate in networks. In Sect. 5.2, inspired by charged system theory in physics, Bi et al. [20] propose a novel influence model, namely charged system influence (CSI) model, capturing how users make decisions among multiple social influences. User behaviors in a social network are affected by multiple factors such as personal interests, social influence, and global trends. The prediction of user behaviors is still in preliminary stages, but has significant potential for understanding temporal user behavior and learning dynamic evolution of social networks. In CSI model, it is the attraction from a specific influence makes a user choose to spread it among multiple influences. Furthermore, an efficient algorithm based on CSI model is presented to predict user behaviors in social networks. Extensive experiments on three real-world datasets demonstrate that our model and algorithm statistically outperform the state-of-the-art methods in terms of prediction accuracy.

5.2   User Behavior Prediction

Years before, researchers believed that human behaviors such as our decision process was mainly taken by rational thoughts. The truth, however, turned out to be much different. Marketers now realize that human beings are quite emotional in nature, and using a social network to propagate information can significantly affect customer decision-making. This observation motivates research community to incorporate social influence as part of factors in decision-making. Furthermore, it raises a new question: since social influence affects individuals' behaviors, is it possible or to what extent that we can predict human behaviors by taking social influence into account?

Previously, it was very difficult to study the behavior prediction problem due to the lack of availability of data. Recently, with the success of many large-scale online social networks, such as Facebook, Twitter, Pinterest, etc., many virtual communities are loosely formed and are often based around some common interests such as forums on a wide variety of issues ranging from product reviews to presidential campaigns. In all of these virtual communities, the change of an individual's emotions, opinions, or behaviors can influence others in positive or negative ways. This propagation of behavior changes has a profound effect on the collective sentiments in social networks. Understanding how the dynamics of influence propagation affects human behaviors in online social networks can provide rich information for applications such as spread of political views, target marketing, rumor blocking, etc.

There have been quite a few related studies have been conducted, for example, dynamic social network analysis [167, 169], social influence analysis [53, 93, 182, 207], and group behavior analysis [176, 186]. The behavior prediction problem addressed in this paper is very different from these works. Dynamic social network analysis is to model how friendships drift over time using a dynamic model [167] or to investigate how different preprocessing decisions and different network forces such as selection and influence affect the modeling of dynamic networks [169]. Social influence analysis either aims to verify the existence of social influence [53, 93] or tries to quantify the strength of the influence [76, 207]. Group behavior analysis intends to study the patterns of user joining different communities [176], or to learn the classification patterns based on the network structure and content information [186], etc.

It is well recognized that users' actions in a social network are influenced by various complex and subtle factors. Multiple influences coexist in social networks and different influences compete or cooperate with each other. For example, although Apple hit the marketing when it launched the iphone5 in 2012 and it has high level on device, Samsung is still strong and has clearly established itself as the dominant Android smartphone maker. Therefore, when it comes to influence propagation in online social networks, it is better to consider mutual effects of multiple influences instead of treating them independently.

In this section, a unified social influence model is built to study various features that may influence users' dynamic behaviors. In a physical charged system, an electric charge will move either further or closer if the electric force from a specific electric field changes. The electric force is related to both the charge's properties and the electric field's properties, which is similar to human behavior when they are exposed to multiple influences. Inspired by the physical phenomena, we build a model based on charged system theory, which describes influences change people's behavior. Take customer's decision for example. Traditionally in influence propagation field, people's buying behavior is defined as "activated" by his friends according to a certain probability, such as recommendation. In contrast, our model believes that people are attracted by the product and choose it on their own initiative. Accordingly, the attraction includes multiple properties such as friends' influence, products' features, and people's own characters.

There are several challenges to build our influence model, namely charged system influence model (CSI). For instance, how to describe the features in social networks using the characters in charged systems, how to define the relationships between different influences, and how to represent an individual's decision among multiple influences. Note that once a single electric object is attracted to move, the global stable status will be broken and other electric charges will move as well. That is similar as the effect of "word of mouth" in marketing field. We employ Coulomb law to simulate the attraction force and use the proposed model to predict whether a person will be attracted by one specific influence. Our main contributions are as follows:

Firstly, we employ physical charged system theory to build an influence model that describes features in social networks and the progress of influence propagation. This model studies how multiple influence spread in a network with interactions between each other. It considers the factors that how people make decisions under multiple influences at a microscopic scale.

Secondly, based on the proposed CSI model, an algorithm is presented which can effectively predict whether an individual will take a target action within a predefined tolerant time. Our framework has consequently allowed us to predict the choice of human behaviors from multiple influences in social networks.

Thirdly, we evaluate our model and algorithm on three real-world social media datasets. The experimental results suggest our model outperforms state-of-the-art methods in human behavior prediction. Moreover, an evaluation based on SVM shows the features proposed in our model are important for predicting human behaviors with multiple influences.

The rest of the section is organized as follows. In Sect. 5.2.1, a brief overview of related work is introduced. Sect. 5.2.2 gives the definition of our problem. In Sect. 5.2.3 an influence model is built which employs the charged system theory. A new approach based on the model is proposed in Sect. 5.2.4. The simulation results are showed in Sect. 5.2.5.

5.2.1   Related Work

There is a large portion of work focused on the single influence source maximization problem. Kempe [93] gave the definition of that problem, which is to find a set of initial set of users in a social network such that from this set the spread of influence in the network can be maximized. Linear threshold (*LT*) model and independent cascade (*IC*) model are two main approaches to formalize this problem. These two models assume that nodes have two states, either activated or inactivated. The node will never change its state after it had been activated. In the former, every activated node contributes some weights to their neighbors and one node will be activated if it received enough weights from all of its neighbors. In the *IC* Model, each activated node has one chance to activate its neighbors with some probability. Wang et al. [198] proposed a *MIA* model and its heuristic algorithm to address the scalability and efficiency issues in large-scale networks.

The study of factors that lead to the influence propagation is also a hot topic in recent years [53, 161]. Domingos et al. [53] computed customers' probability of buying based on Markov random fields. They valued customers' benefit according to their trade history combined with the discount cost of offering to them. Some researchers studied whether there exists influence from friends such that the influence can change people's behavior [75, 87, 164]. Goyal et al. [75] proposed a model that combines the social structure and action logs to predict when a user may perform an action. Their probability based model is under a single influence case and only to test the influence from one user to another user. Bakshy et al. [14] used several attributes of users on Twitter as the features of a regression tree model, to predict the influence of each user. Their model mainly focuses on finding a fair. influence measurement for each individual such that their prediction can provide better targeting strategies in the marketing.

In recent years, there are numerous works studying social influence with multiple influencers. Most of them studied the competitive relationship between two influencers. Some studied this problem using game theory. Bharathi et al. [19] formulated the influence maximization problem as a graph coloring game. Each player selects a set of nodes as seeds to color other nodes with certain probability. The goal of each player is to maximize the number of its colored nodes. They proved for the last player, the greedy algorithm is within a factor $(1 - 1/e)$ of the best response. Goyal et al. [74] developed a game-theoretic framework for the study of influence maximization problem with two competitors. They built a switching-selection model to simulate the influence spreading process. They showed that network structure can interact in dramatic ways with the switching and selection functions at equilibrium. Two competing influences usually applied in health care. Karrer et al. [91] studied the behavior of two mutually exclusive diseases. They derived the phase diagram for the system and discovered the growth rate boundary in the system. Beutel et al. [18] studied the non-mutually competition case and showed that weaker competitor can survive if the cross-immunity satisfies a threshold condition. Their propagation model is based on the *SIS* (susceptible-infected-susceptible) model, in which the infected and self-healing rate are given parameters.

For the influence minimization problem with two competing campaigns, Budak et al. [30] extended the *IC* model to the *Multi-Campaign Independent Cascade Model (MCICM)* and the *Campaign-Oblivious Independent Cascade Model (COICM)*. The former model assumes when two campaigns try to active one node at the same time, "*good information*" takes effect. While in the latter model every edge has the same probability to spread information. They proved the problem of limiting another campaign is NP-hard and submodular for *MCIFCM*. They also studied the problem in the presence of missing the state of some nodes and a prediction algorithm which is based on random spanning trees was proposed.

The two competing influence propagation problems also have a wide use in viral marketing. Carnes et al. [33] studied the influence maximization problem with two exclusive competing products in which each follower has a fixed budget. They proposed two models in which the adopting probability on the links depends on the number of reachable adopters. And they proved their problem is submodular and monotone. Prakash et al. [156] used data from Google, Insights, Facebook, and Myspace to study two competing products spreading over a given network. They used the SIS model under different graph structures to show one virus will completely win another one if its strength is above threshold and greater than another.

For propagation of more than two influences, Pathak et al. [151] proposed a linear threshold model that allows k cascades propagating. It describes the process as Markov chain in which nodes' states can change. Their *StochColor* algorithm discovers the most likely states of the cascades' spreading in a given graph. Myers et al. [138] looked at multiple information diffusing through social networks. They used data on Twitter to learn interactions between information. They built a model to predict the contagion probability. Their experimental results showed stronger contagion has negative effect on unrelated subject matter but has positive effect on high related subject matter.

5.2.2   Behavior Prediction Problem

In this section, after presenting the progress of influence propagation in the view of each individual in the social network, we formally define the targeted problem in this work.

5.2.2.1   Influence Propagation System

We define that the influence propagation system consists of three phases, *Influence Phase*, *Comparison Phase*, and *Action Phase*. At any specific point, one individual was exposed to multiple influences. For example, people can obtain news when they surf on Twitter or get product information when they are in a mall. In *Influence Phase*, an individual's behavior is affected by several factors. Then they will move to the second phase, *Comparison Phase*, in which they analyze, compare, and filter these influences. At a very moment, which here we define as *Action Phase*, they are willing to spread the news or purchase the product. In the marketing, this phase usually happens when the company does the promotion or it is nearly festival.

In particular, the factors in *Influence Phase* are generally classified into the following three parts.

User Interest Bias Personal interests make one person more likely to access one specific influence source. For example, people who love sports are more likely to pay attention to sports news and related products.

Social Structure In many previous works it has been proven that friends' behavior can affect people's behavior. In this paper, our model considers the friends' influence, but the influence spreading does not depend on these friendship links.

Global Attention People are more likely to talk about topics that most people are talking about. For example, when there happens an earthquake most people will pay attention to this event.

The above three factors affect people in the influence phase. We will discuss them further in the following part.

5.2.2.2   Problem Definition

Based on the influence propagation system described above, now we formulate our problem. Given a *social graph* $G = (V, E, T_g)$, where V is the node set representing the users, E is the edge set which represents the social relationship between two users, and T_g refers to the time point when we got the graph snapshot. We also have an *action log ACT*, which contains multiple tuples (u, a, t), indicating that user u takes the action a when $T = t$. Let the set A be the collection of all actions, $A = a_1 \cup a_2 \cup \ldots \cup a_n$. a_i can be any behavior in the social network. It can be

Fig. 5.1 Action prediction with multiple influences

seen as adopting one kind of products under the scenario of marketing. a_i can be seen as joining a group if we consider the problem of maximizing the size of communities. The relationship between two actions can be *competitive, mutualistic,* or *independent*. One person can perform any kind of actions with the time series. Our task is to predict whether an individual will perform one special action. We define the action that we focused on is a^*. We define that once a person takes the action a^*, he or she holds the supportive attitude to a^*, no matter what other actions he or she does after. Since it will take some time for people to perform the action even they have been affected by some influences, we define two tolerant time windows Win_{in} and Win_{act}. We consider people's actions in $t + Win_{act}$ are the results of influences they got in $t - Win_{in}$.

Problem 5.1 (Behavior Prediction with Multiple Influence)   Suppose there exist n different influences spreading in a social network G, which makes each individual has n possible actions $\{a_1, a_2, \ldots, a_n\}$. Let U contains all users who have not taken the action a^* by the time t, our task is to predict whether each user in U will perform a^* within $t + Win_{act}$.

The illustration of this problem is shown in Fig. 5.1.

5.2.3   Modeling Social Influence as Electric Charged System

In this section, we first introduce the charged system in physics. Similar in the aspect of being attracted and movement to the physical model, our *CSI* model represents how an individual is affected by multiple influences and explains why an individual will take actions with these influences.

5.2.3.1   Electric Charged System

In physical charged system, a single charge point creates an electric field which exerts force on other charged objects around without touching. This electric force makes a charge particle move closer or further to another. Coulomb shows that for any two charged points i, j, the electric force between them $\mathbf{F_{ij}}$ is proportional to the product of quantity on the two electric particles q_i, q_j and inversely proportional to the square of separation distance between them [83], as described by Function 5.1.

$$\mathbf{F_{ij}} = h\frac{q_i q_j}{r_{ij}^2},\tag{5.1}$$

where h is the Coulomb constant and r_{ij} is the distance between charges i and j. $\mathbf{F_{ij}}$ is a vector quantity. It appears attractive if the charges are of opposite sign and repulsive otherwise.

Besides a single charged point, a set of charges also create electric field which can be measured by putting a test point charged q_1 at a given position. The resultant force on q_1 equals the vector sum of forces exerted by each individual charges. For example, if two charges are present, then the resultant force exerted by particles q_2 and q_3 on particle q_1 is

$$\mathbf{F_r} = \mathbf{F_{21}} + \mathbf{F_{31}} = kq_1(\frac{q_2}{r_{12}^2}\hat{\mathbf{r_{21}}} + \frac{q_3}{r_{13}^2}\hat{\mathbf{r_{31}}}).\tag{5.2}$$

5.2.3.2   Corresponding Features in Social Influence

Similar to charged particle in physics, each individual has its own characters and will be attracted by the influences in the social network. Now we introduce how to transform features in social networks into variables in electric field theory, which forms the basis of our CSI influence model.

User Vitality (q) An individual in the social network is viewed as charged particles. He may choose any action. For people who have not performed the target action a^*, we assume their electrical polarity is negative. In other words, they hold negative attitude to action a^*. Similarly, if people have taken the action a^*, their electrical polarity is positive and they hold the positive attitude to a^*. Moreover, they will attract other "negative" particles to perform a^*. We allow an individual to perform other actions after it performed a^*, but we still see its electrical polarity as positive though its positive influence to others will decay.

The quantity of a particle q refers to the activity vitality of a user. In the real world, if one person shows up frequently in some social media, he will have more opportunities to be influenced by others and is more likely to affect others as well. Besides, user vitality is a feature which is independent of social structure and topic content. Note that charged quantity is independent of electric field. User vitality (q) is defined as charged quantity property in Function 5.3.

$$q_i = \frac{ActNum_i}{\sum\limits_{i=1}^{n} ActNum_i},\tag{5.3}$$

where $ActNum_i$ is the number of user i's activities and n is the total number of users.

Friends Distance (r) An important variable in Function 5.1 is distance r. The distance r_{ij} reflects the action similarity between user i and j. Note that r_{ij} may vary with time, because users may change their actions in different time slots. Before addressing the problem of how to define friends distance, we first need to know several definitions.

Friends Impact Degree In previous work, the similarity between two friends is used to measure the degree of impaction from a friend. However, in the real world even a person has a lot of close friends, he or she may still hold his or her own option. These "stubborn" people make Max-degree strategy lose efficacy. So when analyzing the influence from friends, we take into account the personal feature, *Friends Impact Degree* α, which measures how easy a person can be influenced by friends' behavior. The computation of α is given in Function 5.4.

$$\alpha_i = \frac{1}{|W(i)|} \sum_{j \in W(i)} \frac{|A_i \cap A_j|}{|A_i|} \qquad (5.4)$$

In Function 5.4, $W(i)$ is the friends set of user i. A_i records all the actions of user i. When an individual has greater α value, it has more actions that are same as its friends' actions, which means it is easier to be influenced by its friends.

Influence Correlation Degree Assume multiple influences spread over the social graph. The relationship between any pair of influences can be *competitive, mutualistic*, or *independent. Independent* is a class of relationship between two influence sources where one influence benefits without affecting the other. It compares with *competitive*, where one benefits while the other is restrained, and *mutualistic*, where both two influence sources benefit. Here, *benefit* refers to influence propagation. We see two influence sources as *mutalistic* if they appear together frequently. If one is always absent when the other is spreading, we consider they are more likely to be *competitive*. We compute the correlation degree between two actions a_x and a_y by Function 5.5.

$$Crr(a_x, a_y) = \frac{|c(a_x) \cap c(a_y)|}{|\min[c(a_x), c(a_y)]|} \qquad (5.5)$$

where $c(a_x)$ is the user set who perform a_x in the action log ACT, i.e. $c(a_x) = \{u | (u, a_x, t) \in ACT\}$. Here we set a low-threshold λ_{neg} and a high-threshold λ_{pos} to separate all actions into three classes. If $Crr(a_x, a_y) > \lambda_{pos}$, we say a_x and a_y are *mutualistic*. If $Crr(a_x, a_y) < \lambda_{neg}$, we say a_x and a_y are *competitive*. Otherwise, we say they are *independent*. For users who take action a_x, if they also take action a_y at the same time, we say $Crr(a_x, a_y) = 1$.

Target Action Correlation Degree Since a^* is the influence that we focused on, we only need to compare other actions with a^* and obtain their correlation degree. Suppose user i takes the action a_x. For convenience we define the i's action correlation degree towards target action a^* by Function 5.6.

$$\delta_i = Crr(a_x, a^*) \tag{5.6}$$

Based on the above variables, the definition of friends distance between user i and user j during a time slot is now described in Function 5.7.

$$r_{ij} = dis_{ij} * |\delta_i - \delta_j|, \tag{5.7}$$

where dis_{ij} means the number of links from i to j. δ_i is the user i's action correlation degree towards a^* during the time slot. Since in experiments we only consider "immediate friends," dis_{ij} should be 1. Therefore, about the definition of friend distance r, in addition to the link distance in the social graph, we also consider the action similarity between two users, which reflects their attitudes difference towards one influence.

5.2.3.3   Expansion Progress and Predication Features

In this part we use electric field theorem to explain the progress of multiple influences propagation. Two prediction features resultant force F_R and displacement S_R will be given.

Resultant Force (F) Suppose at the tth time slot, there are n influences spreading over the social graph G, which corresponds to n kinds of actions. For a user i, it receives the electric force exerted by other users. For users who perform action a_x, where $Crr(a_x, a^*) > \lambda_{pos}$, they have *positive attraction* to user i, because their actions are all closer to target action a^*. They are positive examples for i to perform a^*. For neighbors who perform action a_x, where $Crr(a_x, a^*) < \lambda_{neg}$, they have *negative attraction* to user i, because these people lean to actions that are *competitive* with a^*. For people who perform *independent* actions, their action have no effect to i. To simplify our work, we assume $\lambda_{pos} = \lambda_{neg} = \lambda$, such that there are only *positive attraction* and *negative attraction* in the view of action a^*. As illustrated in Fig. 5.2, at the tth time slot, the user i received *positive force* and *negative force* from its neighbors. The resultant electric force from i's neighbors to user i can be computed by Function 5.8.

$$F_t(i) = \sum_{j \in W(i)} \mathbf{F_{j \to i}} = \alpha_i \sum_{j \in W(i)} (\delta_j - \lambda) \frac{q_i q_j}{r_{ij}^2} \tag{5.8}$$

where α_i reflects how much influence user i accept from their neighbors $W(i)$, which can be computed by Function 5.4. The computation of q_i, q_j can be found in Function 5.3. The computation of δ_j can be found in Function 5.5. The computation of r_{ij_1}, r_{ij_2} can be found in Function 5.7. $\delta_j - \lambda$ guarantees that the force has

Fig. 5.2 Resultant force
analysis

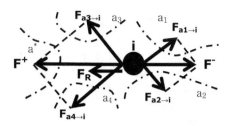

positive value when the neighbor' action is *mutualistic* to a^*, while the force has
negative value when the neighbor' action is *competitive* to a^*. So the resultant
force from neighbors to user i is the difference between the *positive* influence and
negative influence. In general Function 5.8 reflects how many influence regarding
a^* a user can accept from its neighbors, including the user's personal characters,
social structure, and the neighbors' behaviors.

In the real world the influence towards a person will last for a while. As a result,
to estimate whether a user i will perform action a^*, we should consider the total
influence during a period of time, not a single time slot. We separate the time series
into several time slots. We build an *influence window* Win_{in}, which contains k time
slots. For each user i, we calculate the total resultant force during these k time slots
by Function 5.9,

$$F_R(i) = \sum_{m=0}^{k-1} e^{-\gamma m} F_{t-m}(i) , \qquad (5.9)$$

where γ is the *decay constant*. $e^{-\gamma m}$ shows that the influence towards user i decays
exponentially with time, which guarantees that the latest state has stronger influence
to an individual. It shows that user i was affected by positive influence during last
k time slots when $F_R(i) \geq 0$, which means i is more likely to perform a^* than
users who have negative F_R. F_R will be considered as a prediction feature in the
algorithm part.

State Displacement (S) In the physics, a negative electric particle will move closer
to positive electric particles due to the attractive resultant force. Similarly, after
exposed to several influences in the social graph, it is possible for some users to
perform the target action a^* when they obtain enough *positive* resultant force
from their neighbors. However, whether they will perform a^* is uncertain. We see
performing the target action a^* as the goal state. When a user i takes the action a_x,
its *influence correlation degree* to a^*, δ_i, reflects how far to the goal state. We define
$S(i)$ as the *displacement* between user i's current state and the goal state, which can
be computed by Function 5.10,

$$S_t(i) = \mu_1(1 - \delta_i) + \mu_2 \alpha_i \sum_{j \in W(i)} \frac{1 - \delta_j}{r_{ij}} , \qquad (5.10)$$

where δ_i and δ_j can be computed by Function 5.5. r_{ij} can be computed by Function 5.7.

The Function 5.10 shows that the displacement has two parts. One is the user's own distance to the goal action, i.e., $1 - \delta_i$. If user i's current action is closer to a^*, $S(i)$ is smaller which means i is more likely to perform a^*. The other part is the distance between the actions of i's neighbors and the target action a^*. If people around user i all perform actions that are competitive with a^*, it will take more effect for i to achieve the goal. Otherwise user i will have greater possibility to perform a^*. μ_1 and μ_2 are the parameters which adjust the weight of these two parts in the state displacement. In previous work, the influence must spread through the links in the social graph, which means if one person adopts an item, there at least exists one neighbor who already adopted this item. However, in the real world it is possible that one person will try a new item even the neighbors have not adopted the new item. In fact, when μ_2 or α_i in Function 5.10 equals to 0, the neighbors have no effect on user i's behavior.

Similar to F_R, we define the state displacement during the influence window Win_{in} by Function 5.11.

$$S_R(i) = \sum_{m=0}^{k-1} e^{-\gamma m} S_{t-m}(i) \tag{5.11}$$

Generally, the feature *State Displacement* looses the constraint that influence must spread from an *activated* user to an *unactivated* user. S_R is also considered as a feature for predicting user's action in the following part.

5.2.4   Algorithms

In this section, we want to use the resultant force from neighbors F_R and the state displacement S_R as features to predict whether a user will take the target action a^* during the action window Win_{act}. Win_{act} includes n time slots after the current time slot. We propose an algorithm which is named **APMI** (*Action Prediction under Multiple Influences*) to solve this problem.

We have the sample of observed data from *influence window* Win_{in}, where each datapoint includes user's features and its action label. We define user i's label $y_i = 1$ if its action is a^*. Otherwise, its label $y_i = 0$. Since *Gaussian distributions* are widely used in social sciences, we assume the observed features we obtained have the multivariate Gaussian distributions of dimension d. The main idea of **APMI** is as follows: (1) Create a mixture of two multivariate Gaussian distribution functions based on users' label; (2) Estimate the parameters of each multivariate Gaussian distribution function using **maximum likelihood estimation (MLE)**; (3) Predict user's action by comparing the probability of performing a^* and the probability of not performing a^*.

Density Estimator We are given the sample of observed features $X =$ $\{X_1, X_2, \ldots, X_k\}$, where k is the number of users in the sample. Each observation X_i is a vector which has d components. Assume each component generates data from a Gaussian distribution; therefore, X has the multivariate Gaussian distribution of d dimensions. We are also given the action labels $Y = \{y_1, y_2, \ldots, y_k\}$ that correspond to the observations X. Based on the labels we have a mixture of two multivariate Gaussian distributions with mean μ_i and covariance matrix Σ_i. One contains observations whose labels equal to 1 and the other contains observations whose labels equal to 0, i.e.,

$$\begin{cases} X|(Y = 1) \sim N_d(\mu_1, \Sigma_1) \\ X|(Y = 0) \sim N_d(\mu_0, \Sigma_0) \end{cases}$$

From the given label data, we can compute label 1's prior probability π_1 by counting the number of users who took the action a^*.

$$\begin{cases} P(Y = 1) = \pi_1 \\ P(Y = 0) = \pi_0 = 1 - \pi_1 \end{cases}$$

Since the observations are independent and identically distributed, the probability of obtaining observations X and labels Y can be computed by Function 5.12.

$$P(X, Y|\phi) = \prod_{i=1}^{k} \sum_{j=0}^{1} \mathbf{1}(y_i = j)\pi_j f(X_i; \mu_j, \Sigma_j) \tag{5.12}$$

where f is the probability density function of the d dimensions multivariate Gaussian distribution. ϕ is the set of unknown parameters in f, i.e., $\phi = (\mu_1, \mu_0, \Sigma_1, \Sigma_0)$. $\mathbf{1}(\cdot)$ is the label indicator function.

$$L(\phi; X, Y) = exp\{\sum_{i=1}^{k} \sum_{j=0}^{1} \mathbf{1}(y_i = j)[log\pi_j - \frac{1}{2} \log |\Sigma_j|$$
$$\tag{5.13}$$
$$- \frac{1}{2}(X_i - \mu_j)^\top \Sigma_j^{-1}(X_i - \mu_j) - \frac{d}{2} \log 2\pi]\}$$

Function 5.13 is the exponential family form of Function 5.12. Since log is a monotonically increasing function, maximizing the likelihood of Function 5.13 is the same as maximizing its log form. Function 5.14 shows the method of *MLE* that selects the set of model parameters which maximize the likelihood function.

$$(\mu_j, \Sigma_j) = \arg\max_{\mu_j, \Sigma_j} \log L(\phi; X, Y)$$

$$= \arg\max_{\mu_j, \Sigma_j} \sum_{i=1}^{k} \mathbf{1}(y_i = j)\{-\frac{1}{2}\log|\Sigma_j| \qquad (5.14)$$

$$-\frac{1}{2}(X_i - \mu_j)^\top \Sigma_j^{-1}(X_i - \mu_j)\}$$

To estimate ϕ, we take the derivatives of $\log L(\phi; X, Y)$ with respect to μ_j and Σ_j. i.e., Function 5.15.

$$\begin{cases} \frac{\partial \log L(\phi; X, Y)}{\partial \mu_j} = 0 \\ \frac{\partial \log L(\phi; X, Y)}{\partial \Sigma_j} = 0 \end{cases} \qquad (5.15)$$

By solving Function 5.15, we get Function 5.16.

$$\begin{cases} \mu_j = \frac{\sum_{i=1}^{k} \mathbf{1}(y_i = j) X_i}{\sum_{i=1}^{k} \mathbf{1}(y_i = j)} \\ \Sigma_j = \frac{\sum_{i=1}^{k} \mathbf{1}(y_i = j)(X_i - \mu_j)(X_i - \mu_j)^\top}{\sum_{i=1}^{k} \mathbf{1}(y_i = j)} \end{cases} \qquad (5.16)$$

Let $\mathbf{1}(y_i = 1) = 1$ when the user performs a^*, otherwise $\mathbf{1}(y_i = 0) = 0$. Since the labels for each user's actions during Win_{in} are observed, we can compute each parameter in the Gaussian function by Function 5.16.

Action Prediction Now we have the probability density function for each prediction feature without unknown parameters. We can compute all labels' probability under any observation X_i. Function 5.17 shows that the conditional distribution of label Y_i is determined by Bayes theorem.

$$P(y_i = j | X_i = x_i) = \frac{\pi_j f(x_i; \mu_j, \Sigma_j)}{\sum_{j=0}^{1} \pi_j f(x_i; \mu_j, \Sigma_j)} \qquad (5.17)$$

If $P(y_i = 1 | X_i = x_i) \geq P(y_i = 0 | X_i = x_i)$, we predict user i will take the target action a^*. Otherwise, we predict i will take other actions. Algorithm 1 shows the whole progress of prediction. Step 1 to step 8 are forming the Gaussian function for prediction features. Step 9 to step 14 use Bayes theorem to predict the users' action.

5.2.5 Experiment

In this section, we discuss in detail the experiments that we predict users' behavior using the proposed *CSI* model and *APMI* algorithm on three large real social media datasets.

Algorithm 1 APMI algorithm

Input: $G = (V, E)$ social graph
ACT action log
a^* target action
Win_{in} influence window
Output:action label y_i for each $i \in U/U^*$.
   $U \leftarrow \{user|\ user$ is activated in $Win_{in}\}$;
   $U^* \leftarrow \{user|\ user$ performed a^* in $Win_{in}\}$;
   for all $i \in U/U^*$ **do**
      Compute $F_R(i), S_R(i)$ by Function 5.9 and 5.11;
      Create the feature vector X_i with $F_R(i)$ and $S_R(i)$;
   end for
   $P(Y = 1) \leftarrow \frac{|U^*|}{|U|}$;
   $P(Y = 0) \leftarrow 1 - P(Y = 1)$;
   Compute μ and Σ for feature vector X by Function 5.16;
   Compute the probability of i performs a^* and that of i does not perform a^* by Function 5.17;
   if $\frac{P(y_i=1|X_i=x_i)}{P(y_i=0|X_i=x_i)} \geq 1$ **then**
      $y_i = 1$;
   else
      $y_i = 0$;
   end if
   return action label y_i for all $i \in U/U^*$.

5.2.5.1   Experiment Setup

1. *Experimental Datasets*: We carried out all our experiments on three different social media datasets:

   - **Digg** is a social news website that allows users to vote the posted stories. This dataset includes 279,631 users and 2,251,171 directed friendship, where the friendship refers to one user is watching another's activates. There also exit an action log which record 3,018,197 votes on 3553 popular stories made by 139,409 distinct users during a month.
   - **Flixster** is an online social community that allows users to share movie reviews and ratings with their friends. This dataset contains a social graph with 922,212 users and 7,058,819 friendship links. It also contains a rating log which has 8,196,077 votes for 48,794 movies. For convenience, we ignore the exact score in the rating log. Instead we think once a user rates for a movie, he or she has made an action for the influence spreading.
   - **Douban** is a Chinese website allowing users to record information and create content related to film, books, music and so forth. This dataset concludes 4778 users and 104,799 directed links. The directed link refers to who follows who in Douban. There is also a review log which records 2,549,523 reviews towards 232,160 books.

2. *Evaluated Methods*: Note that in our *APMI* algorithm, prediction feature vector X may have d dimensions. We first employ *resultant force* (F_R) and *state*

displacement (S_R) as feature separately. Then we use these two features together to evaluate our algorithm. For baseline, we use the idea of three methods mentioned in [138], i.e., UB, AC, and EC.

- **Prior Probability + User Bias (UB)**. The independent cascade model assumes that the spreading of one influence is independent with other influences. So the prior probability of one action spreading is related to its popularity, which can be computed simply by counting the number of users who have taken this action. User bias refers to how frequently a user takes an action, which is the same meaning as *user vitality* in Function 5.3. Thus, $popularity(a^*) = \frac{|U^*|}{|U|}$ and q_i form the feature vector X in this baseline method, where U is the total users set and U^* is the set that contains users who have taken action a^*.
- **Best Action Correlation (AC)**. Including prior probability $popularity(a^*)$ and user bias q_i, this baseline method also considers the most positive action in the past for user i. Here *most positive action* refers to action which has the *highest influence correlation* to a^*. Thus, this method has three attributes to form its prediction feature vector.
- **Exposure Curves (EC)**. The idea of exposure curves [138] is that a probability of be activated depends on the number of times when the user was exposed to the influence. In our datasets, the opportunity of being exposed to an influence is when the neighbors perform the corresponding action. Thus, EC baseline counts the number of users who have taken *positive* actions and takes the number as a prediction feature. Here, *positive* means the influence correlation to a^* is greater than the threshold λ.
- **APMI-F (F)**. This method only considers F_R in Function 5.9 as the prediction feature.
- **APMI-S (S)**. This method only considers S_R in Function 5.11 as the prediction feature.
- **APMI-F&S (F&S)**. This method considers F_R and S_R as the prediction features.

3. *Default Parameters*: The threshold λ determines whether an action is *positive influence* or *negative influence*. Its default value is set as 0.4. But several experiments by tuning the λ are presented in the following. The influence decay constant γ is set as 1. Weight parameters μ_1 and μ_2 in Function 5.10 are set as 0.5, which means we treat the friends' state and the user's own state equally.
4. *Measurement*: We use three criterions: **precision**, **recall**, and **F-score** to evaluate the prediction task. As mentioned before, we set a tolerate window Win_{act}. If we predict a user will perform a^* and he or she does take the action a^* during Win_{act}, we see this prediction is true positive.

5.2.5.2   Experimental Results

In this subsection we report the performance of different methods for predicting
whether the user will take the target action. The action logs in the three datasets
record users' behavior according to the time series. Since the Flixster dataset has
no exact time stamp with each action, we divide all the three data sets into ten
units, $\{W_1, W_2, \ldots, W_{10}\}$, such that each unit W_i has substantially equal number of
actions. Note that the divided data still keeps chronological order, i.e., the actions
in W_i take place after the actions in W_{i-1}. In the following part, we will show the
results from three aspects: (1) How the action tolerant window Win_{act} affects the
prediction performance; (2) How the influence window Win_{in} affects the prediction
performance; (3) Whether choosing different actions will affect the performance; (4)
How the parameter λ affects the results.

1. *Effect of* Win_{act}. We randomly choose one action as the target action. Set the first
   six data units as the influence window and change the size of the action tolerant
   window Win_{act} from 1 to 4. The results on Digg dataset are shown in Fig. 5.3,
   where x axis represents the recall (%) and y axis represents the precision (%).
   As expected, each method's recall and precision value raise with the enlargement
   of Win_{act}. The lager tolerant action window improves the prediction accuracy.
   For example, the precision of UB is about 18% when Win_{act} is 1 while it
   raises to 28% when Win_{act} is 4. Compare to other three baselines, our F and
   S methods perform better in both recall and precision value. For example, in
   Fig. 5.3d, compare to UB, F performed almost 50% better in the precision value
   while S performed almost 67% better in the recall value. The remarkable result
   is that $F\&S$ significantly outperform F and S, which means the features in our
   CSI model can improve the prediction performance.
2. *Effect of* Win_{in}. To measure the effect of the influence window, we change
   Win_{in} size from 6 to 9 and set the first after unit as Win_{act}. The results in Fig. 5.4
   show that the enlargement of Win_{in} improves the prediction performance, which
   means our model is more fitting the datasets when the training set is larger.
   The performance of F, S, and $F\&S$ is still better than those of other three
   methods. Though all methods have improved their precision or recall value,

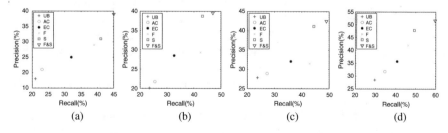

Fig. 5.3 Effect of action tolerant window size Win_{act} on Digg. (**a**) $Win_{act} = 1$. (**b**) $Win_{act} = 2$.
(**c**) $Win_{act} = 3$. (**d**) $Win_{act} = 4$

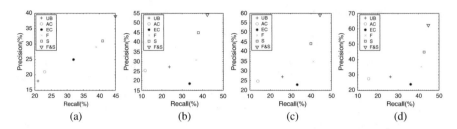

Fig. 5.4 Effect of influence window size Win_{in} on Digg. (**a**) $Win_{in} = 6$. (**b**) $Win_{in} = 7$. (**c**) $Win_{in} = 8$. (**d**) $Win_{in} = 9$

Table 5.1 Average F-scores (%) using six algorithms and ten target actions on three datasets with different Win_{in} and Win_{act}

Data set	Model	T6W1	T6W2	T6W3	T6W4	T7W1	T8W1	T9W1	WTL
Digg	BU	19.38	25.18	25.84	27.02	26.23	26.36	28.11	0/0/7
	AC	21.95	29.94	30.66	35.15	31.68	32.55	33.91	0/0/7
	EC	28.07	35.08	36.32	38.01	33.63	37.10	36.38	0/0/7
	F	33.26	42.23	46.28	47.23	42.33	46.40	48.67	0/0/7
	S	35.31	43.38	44.61	45.82	43.32	44.18	46.50	0/0/7
	F&S	**41.79**	**53.10**	**57.03**	**60.00**	**54.07**	**57.11**	**59.39**	**7/0/0**
Flixster	BU	21.77	28.75	30.21	31.39	27.75	29.92	30.98	0/0/7
	AC	24.95	34.90	35.35	40.62	34.14	34.53	40.52	0/0/7
	EC	33.08	41.14	43.08	49.82	40.78	43.02	49.30	0/0/7
	F	38.99	50.03	54.97	56.24	49.85	54.76	55.33	0/0/7
	S	40.74	51.47	52.64	64.55	51.00	51.73	64.44	0/0/7
	F&S	**49.15**	**62.67**	**67.77**	**70.65**	**61.94**	**67.21**	**70.47**	**7/0/0**
Douban	BU	20.08	26.16	28.41	34.95	26.81	27.98	33.99	0/0/7
	AC	19.83	30.35	33.52	35.51	31.84	33.33	35.56	0/0/7
	EC	31.37	43.82	48.51	52.03	42.62	43.57	50.82	0/0/7
	F	34.16	43.11	50.57	52.72	44.75	49.38	50.95	0/0/7
	S	36.96	44.28	47.28	53.34	45.75	46.39	54.90	0/0/7
	F&S	**42.18**	**57.45**	**60.68**	**65.42**	**56.89**	**61.37**	**64.81**	**7/0/0**

F&S's improvement with Win_{in} is remarkable. When Win_{in} size enlarges to 9, its precision improves to 32.6% comparing with its precision when $Win_{in} = 6$. Due to limitations of space, Figs. 5.3 and 5.4 only show the precision and recall results on Digg. But our algorithms still perform better than baselines on other two datasets, which can be seen in Table 5.1.

3. *Different target action.* A set of experiments were repeated by running all methods with different ten target actions on three datasets with different sizes of Win_{in} and Win_{act}. These 10 actions are randomly chosen from the top 20 popular actions. Final results with error bar are presented in Figs. 5.5 and 5.6, which show that F-score is increasing with the growth of the size of Win_{act} and Win_{in}. Note that EC performs better than AC and UB on all datasets

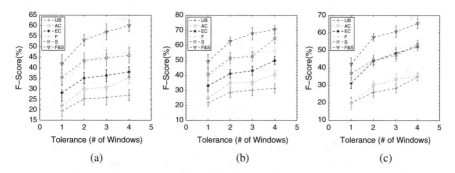

Fig. 5.5 F-score of different target actions with changing action tolerant window size on three datasets. (**a**) Digg $Win_{in} = 6$. (**b**) Flixster $Win_{in} = 6$. (**c**) Douban $Win_{in} = 6$

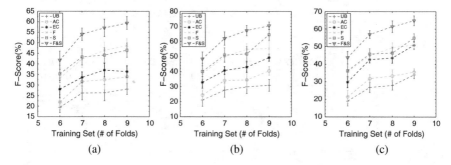

Fig. 5.6 F-score of different target actions with changing influence window size on three datasets. (**a**) Digg $Win_{act} = 1$. (**b**) Flixster $Win_{act} = 1$. (**c**) Douban $Win_{act} = 1$

with different Win_{act} and Win_{in}. This is because EC takes into account the neighbors' influence, while AC and UB only consider about the user's personal features. Compared to UB, AC has more prediction features, which probably explains why AC has higher F-score value than UB. Similarly, though F and S both perform better than EC, $F\&S$ which considers these two features together has outstanding performance than other methods. For example, on the Flixster dataset, $F\&S$'s F-score has achieved 70% when the $Win_{in} = 6$ and $Win_{act} = 4$, while EC which has the best performance in baselines, only has achieved 32%.

Tables 5.1 and 5.2 list the results about the average and standard deviations of the F-scores (%) on three data sets with different Win_{in}, Win_{act}, and target action using six different methods. The notation $T6W1$ means the Win_{in} size is 6 and Win_{act} size is 1. The t-test results under the significance level of 0.05 are summarized in the "WTL" column, where "WTL" (Win/Tie/Loss) represents the numbers of datasets whose corresponding algorithm has higher/equal/lower F-score than others'. From this item we can see that $F\&S$ achieves best results with the varying of Win_{in}, Win_{act}, and target action, whose performance over other methods are statistically significant. For instance, on the Flixster dataset,

Table 5.2 Standard deviations of the F-scores (%) using six algorithms and ten target actions on three datasets with different Win_{in} and Win_{act}

Data set	Model	T6W1	T6W2	T6W3	T6W4	T7W1	T8W1	T9W1	WTL
Digg	BU	±2.59	±3.02	±3.62	±2.68	±3.33	±3.87	±2.56	0/0/7
	AC	±3.91	±2.80	±4.31	±2.86	±3.09	±4.76	±2.68	0/0/7
	EC	±3.97	±2.67	±2.90	±2.46	±2.61	±3.23	±2.72	0/0/7
	F	±3.55	±3.79	±3.48	±3.21	±4.15	±3.97	±3.22	0/0/7
	S	±3.70	±3.71	±3.78	±3.68	±3.80	±3.43	±3.38	0/0/7
	F&S	**±4.38**	**±2.72**	**±3.85**	**±2.78**	**±2.97**	**±4.17**	**±3.07**	**7/0/0**
Flixster	BU	±2.59	±3.42	±3.69	±3.03	±4.72	±4.42	±3.95	0/0/7
	AC	±4.15	±2.88	±4.06	±3.02	±4.22	±5.86	±2.89	0/0/7
	EC	±3.56	±2.80	±3.06	±2.68	±2.60	±3.46	±2.31	0/0/7
	F	±3.94	±4.27	±3.75	±3.29	±2.93	±3.70	±4.63	0/0/7
	S	±4.13	±3.79	±3.29	±3.30	±4.79	±4.78	±3.73	0/0/7
	F&S	**±4.74**	**±2.70**	**±4.19**	**±2.49**	**±4.46**	**±3.49**	**±2.57**	**7/0/0**
Douban	BU	±3.21	±3.17	±2.84	±2.50	±2.87	±2.60	±2.19	0/0/7
	AC	±3.72	±2.42	±3.57	±1.94	±2.30	±3.50	±1.91	0/0/7
	EC	±2.91	±1.65	±2.91	±1.82	±1.90	±2.95	±2.10	0/0/7
	F	±2.84	±3.33	±2.39	±2.72	±2.85	±2.86	±2.69	0/0/7
	S	±3.55	±3.31	±3.78	±2.00	±3.01	±3.29	±2.43	0/0/7
	F&S	**±4.00**	**±1.73**	**±3.56**	**±3.08**	**±2.21**	**±3.37**	**±2.79**	**7/0/0**

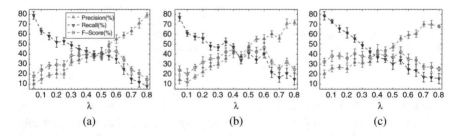

Fig. 5.7 Average precision/recall/F-score of different λs on three datasets. (**a**) Digg. (**b**) Flixster. (**c**) Douban

when $T = 6$ and $W = 4$, the best performance of baseline is 49.84%, while $F\&S$ achieves 70.65%.

4. *Tuning of parameter* λ. In Function 5.8, parameter λ decides whether the influence from neighbors' actions is positive towards the target action. It is a threshold for assessing the effect of each action's *target action correlation degree* δ_i. To evaluate how λ affects the prediction accuracy of F_R feature, we construct a set of experiments by tuning λ from 0 to 0.8 on three datasets with ten different target actions. Figure 5.7 shows the results of the experiments where Win_{in} concludes the first six data units and Win_{act} is the seventh data unit. From Fig. 5.7, we can observe that (1) precision value increases with the growth of λ. It is small at the beginning, because when λ is small, our model thinks most

neighbors' actions have the positive influence to a user. So our algorithm will predict more people to perform the target action. The predictions of people who are expected to perform the target action, but actually not, decrease the precision; (2) Recall value decreases with the growth of λ. The reason of that trend is that when λ value is high, our model thinks most of neighbors have negative influence to the user, which underestimates their positive attitude towards the target action. As a result, it will miss some people who are actually affected by their neighbors and performed the target action. Since F-score is a balance between precision and recall, it will decrease when one of them has a low value; (3) On all datasets, the curves of three measurements intersect when λ is around 4.5. Recall that λ is related to the correlation degree to the target action. When the correlation degree is greater than 0.5, the very action has more than half opportunity to appear together with the target action, which means that action indicates the appearance of the target action in some degree. Thus the experimental results suggest that *Influence Correlation Degree* can reasonably represent the relationship between two different influences.

5.2.5.3   Evaluation of Prediction Features

In Sect. 5.2.3 we provide two features F_R and S_R, and use them in our APMI algorithm for predicting human behavior which show remarkable results in our experiments. In this part we want to compare these two features with other common prediction features in social networks and evaluate their performance under the same algorithm. Guyon et al. [81] used the weights of a linear support vector machines (SVMs) to produce a feature ranking. Their SVM-RFE algorithm and experimental results show that, for a feature vector **w**, the larger weight w_i is, the corresponding ith feature plays a more important role in decision function.

Based on Guyon's work, we use the weights of SVM to evaluate the effect of several factors on predicting human behavior under multiple influences. We choose the following ten factors to form the feature vector for SVM: (1) F_R, (2) S_R, (3) number of actions (influences), (4) number of users, (5) number of users who perform the target action, (6) number of actions a user performed, (7) correlation degree between the current action and the target action, (8) number of user's neighbors, (9) popular ratio of the target action, and (10) exposure times of a user to the target action (number of neighbors who have already took the target action).

We use the SVM to do the prediction on the three datasets. For each dataset we choose ten different target actions and changing the training set from 6 units to 9 units. Then we rank the weights of the ten features above and obtain the top five features with higher weights. The results are shown in Table 5.3, where *Ratio* is the times of appearing in top five divide the total number of SVM's training. Obviously, our F_R and S_R have the highest probability to appear in the top five, which proves that the features generated by our model are important for the prediction. *# of actions* is also important because the more activities a person join, the bigger opportunity

Table 5.3 Feature ranking

Feature	F	S	# of actions	# of executors	# of friends
Ratio	97.25	91.69	81.73	69.92	57.33

to access the influence, which is also defined in our model by Function 5.3. *# of executors* refers to the number of users who performed the target action. It ranks at the fourth place, which means the popularity of the influence will affect its propagation. The fifth important feature is the number of user's neighbors, which again proves the power of "word of mouth."

5.3   Conclusion

In this chapter, we discussed a complicated case of information propagation when multiple social influences propagate in social networks. In Sect. 5.2, the work advances the state of the art in modeling single influence propagation. We propose a novel influence model that studies propagation with interacting of multiple influences. Moreover, inspired by the electric attraction in physics, our model considers the propagation factors in microscopic way. It looses the constraints that influence must spread via links in the social graph. An algorithm based on this model effectively learned features to predict human behavior. Our comprehensive experiments have demonstrated the effectiveness and accuracy of our proposed model and algorithm. In the future, we will deal with more features in social networks that can affect human behaviors.

Bibliography

1. L.A. Adamic, O. Buyukkokten, E. Adar, A social network caught in the web. First Monday **8**, 6 (2003)
2. A. Agaskar, Y.M. Lu, A fast monte carlo algorithm for source localization on graphs, in *SPIE Optical Engineering and Applications* (2013)
3. R. Albert, A.L. Barabasi, Statistical mechanics of complex networks. Rev. Mod. Phys. **74**(1), 47–97 (2002)
4. R. Albert, H. Jeong, A.L. Barabasi, The diameter of the world wide web. Nature **401**, 130 (1999)
5. N. Alon, M. Feldman, A.D. Procaccia, M. Tennenholtz, A note on competitive diffusion through social networks. Inf. Process. Lett. **110**(6), 221–225 (2010)
6. L.A.N. Amaral, A. Scala, M. Barthelemy, H.E. Stanley, Classes of small-world networks. Proc. Natl. Acad. Sci. (PNAS) **97**, 11149–11152 (2000)
7. A. Anagnostopoulos, R. Kumar, M. Mahdian, Influence and correlation in social networks, in *Proceeding of the 14th ACM SIGKDD International Conference on Knowledge Discovery and Data Mining (KDD)*, vol. 0 (2008), pp. 7–15
8. R.M. Anderson, R.M. May, *Infectious Diseases of Humans: Dynamics and Control* (Oxford University Press, USA, New York, 1992)
9. N. Antulov-Fantulin, A. Lancic, T. Smuc, H. Stefancic, M. Sikic, Identification of patient zero in static and temporal networks: robustness and limitations. Phys. Rev. Lett. **114**(24), 248701 (2015)
10. S. Aral, D. Walker, Identifying influential and susceptible members of social networks. Science **337**, 337–341 (2012)
11. S. Aral, L. Muchnika, A. Sundararajan, Distinguishing influence-based contagion from homophily-driven diffusion in dynamic networks. Proc. Natl. Acad. Sci. **106**(51), 21544–21549 (2009)
12. L. Backstrom, D. Huttenlocher, J. Kleinberg, X. Lan, Group formation in large social networks: membership, growth, and evolution, in *Proceedings of the 12th ACM SIGKDD International Conference on Knowledge Discovery and Data Mining (KDD), Philadelphia, PA* (2006)
13. L. Backstrom, R. Kumar, C. Marlow, J. Novak, A. Tomkins, Preferential behavior in online groups, in *Proceedings of the International Conference on Web Search and Web Data Mining (WSDM)* (2008), pp. 117–128
14. E. Bakshy, J.M. Hofman, W.A. Mason, D.J. Watts, Everyone's an influencer: quantifying influence on twitter, in *Proceedings of the Fourth ACM International Conference on Web Search and Data Mining* (ACM, New York, 2011), pp. 65–74

15. A.L. Barabasi, R. Albert, Emergence of scaling in random networks. Science **286**, 509–512 (1999)
16. F.M. Bass, A new product growth for model consumer durables. Manag. Sci. **15**(5), 215–227 (1969)
17. P. Berman, B. DasGupta, M. Kao, Tight approximability results for test set problems in bioinformatics. J. Comput. Syst. Sci. **71**, 145–162 (2005)
18. A. Beutel, B.A. Prakash, R. Rosenfeld, C. Faloutsos, Interacting viruses in networks: can both survive? in *Proceedings of the 18th ACM SIGKDD International Conference on Knowledge Discovery and Data Mining* (ACM, New York, 2012), pp. 426–434
19. S. Bharathi, D. Kempe, M. Salek, Competitive influence maximization in social networks. Internet Netw. Econ. **4858**, 306–311 (2007)
20. Y. Bi, W. Wu, Y. Zhu, Csi: charged system influence model for human behavior prediction, in *ICDM* (2013), pp. 31–40
21. V.D. Blondel, J.-L. Guillaume, R. Lambiotte, E. Lefebvre, Fast unfolding of community hierarchies in large networks. J. Stat. Mech. **2008**, P10008 (2008)
22. R.M. Bond, C.J. Fariss, J.J. Jones, A.D.I. Kramer, C. Marlow, J.E. Settle, J.H. Fowler, A 61-million-person experiment in social influence and political mobilization. Nature **489**, 295–298 (2012)
23. P. Bordia, N. DiFonzo, Problem solving in social interactions on the internet: Rumor as social cognition. Soc. Psychol. Q. **67**(1), 33–49 (2004)
24. S.P. Borgatti, M.G. Everett, A graph-theoretic perspective on centrality. Soc. Netw. **28**(4), 466–484 (2006)
25. C. Borgs, M. Brautbar, J. Chayes, B. Lucier, Maximizing social influence in nearly optimal time, in *Proceedings of the Twenty-fifth Annual ACM-SIAM Symposium on Discrete Algorithms* (2014), pp. 946–957
26. A. Borodin, Y. Filmus, J. Oren, Threshold models for competitive influence in social networks, in *WINE* (2010), pp. 539–550
27. V. Braitenberg, A. Schuz, *Anatomy of a Cortex: Statistics and Geometry* (Springer, Berlin, 1991)
28. U. Brandes, A faster algorithm for betweenness centrality. J. Math. Sociol. **25**(2), 163–177 (2001)
29. A. Broder, R. Kumar, F. Maghoul, P. Raghavan, S. Rajagopalan, R. Stata, A. Tomkins, J. Wiener, Graph structure in the web: Experiments and models, in *Proceedings of the 9th International World Wide Web Conference (WWW)* (2000)
30. C. Budak, D. Agrawal, A.E. Abbadi, Limiting the spread of misinformation in social networks. *WWW* (2011), pp. 665–674
31. R.S. Burt, *Structural Holes: The Social Structure of Competition* (Harvard University Press, Cambridge, 1992)
32. J.T. Cacioppo, J.H. Fowler, N.A. Christakis, Alone in the crowd: the structure and spread of loneliness in a large social network. SSRN eLibrary (2008)
33. T. Carnes, C. Nagarajan, S.M. Wild, A.V. Zuylen, Maximizing influence in a competitive social network: a follower's perspective, in *Proceedings of the 9th International Conference on Electronic Commerce (ICEC)* (2007), pp. 351–360
34. C. Castillo, M. Mendoza, B. Poblete, Information credibility on twitter, in *WWW* (2011), pp. 675–684
35. D. Centola, The spread of behavior in an online social network experiment. Science **329**(5996), 1194–1197 (2010)
36. M. Cha, H. Haddadi, F. Benevenuto, K. Gummadi, Measuring user influence in twitter: the million follower fallacy, in *Proceedings of the 4th International Conference on Weblogs and Social Media* (2010)
37. W. Chen, Y. Wang, S. Yang, Efficient influence maximization in social networks, in *KDD* (2009), pp. 199–208
38. W. Chen, C. Wang, Y. Wang, Scalable influence maximization for prevalent viral marketing in large-scale social networks, in *KDD* (2010), pp. 1029–1038

39. W. Chen, Y. Yuan, L. Zhang, Scalable influence maximization in social networks under the linear threshold model, in *ICDM* (2010), pp. 88–97
40. W. Chen, A. Collins, R. Cummings, T. Ke, Z. Liu, D. Rincn, X. Sun, Y. Wang, W. Wei, Y. Yuan, Influence maximization in social networks when negative opinions may emerge and propagate, in *SDM* (2011), pp. 379–390
41. W. Chen, W. Lu, N. Zhang, Time-critical influence maximization in social networks with time-delayed diffusion process, in *AAAI* (2012), pp. 1–5
42. H. Chen, R.H.L. Chiang, V.C. Storey, Business intelligence and analytics: from big data to big impact. MIS Q. **36**(4), 1165–1188 (2012)
43. N.A. Christakis, J.H. Fowler, The spread of obesity in a large social network over 32 years. N. Engl. J. Med. **357**(4), 370–379 (2007)
44. N.A. Christakis, J.H. Fowler, The collective dynamics of smoking in a large social network. N. Engl. J. Med. **358**(21), 2249–2258 (2008)
45. C.-T. Chu, S.K. Kim, Y.-A. Lin, Y. Yu, G.R. Bradski, A.Y. Ng, K. Olukotun, Map-reduce for machine learning on multicore, in *Proceedings of the 19th Neural Information Processing Systems (NIPS)* (2006), pp. 281–288
46. R.B. Cialdini, N.J. Goldstein, Social influence: compliance and conformity. Annu. Rev. Psychol. **55**, 591–621 (2004)
47. M. Corcoran, Death by cli plunge, with a push from twitter. New York Times (2009)
48. D. Crandall, D. Cosley, D. Huttenlocher, J. Kleinberg, S. Suri, Feedback effects between similarity and social influence in online communities, in *Proceeding of the 14th ACM SIGKDD International Conference on Knowledge Discovery and Data Mining (KDD)* (2008), pp. 160–168
49. A. Das, M. Datar, A. Garg, S. Rajaram, Google news personalization: scalable online collaborative filtering, in *Proceeding of the 16th International Conference on World Wide Web (WWW)* (2007), pp. 271–280
50. N. DiFonzo, P. Bordia, Rumor, gossip, and urban legend. Diogenes **54**, 19–35 (2007)
51. P. Dodds, D. Watts, Universal behavior in a generalized model of contagion. Phys. Rev. Lett. **92**, 218–701 (2004)
52. B. Doerr, M. Fouz, T. Friedrich, Social networks spread rumors in sublogarithmic time, in *Proceedings of the 43rd Annual ACM Symposium on Theory of Computing* (2011), pp. 21–30
53. P. Domingos, M. Richardson, Mining the network value of customers, in *Proceedings of the Seventh ACM SIGKDD International Conference on Knowledge Discovery and Data Mining* (ACM, New York, 2001), pp. 57–66
54. W. Dong, W. Zhang, C.W. Tan, Rooting out the rumor culprit from suspects. *IEEE International Symposium on Information Theory (ISIT)*, pp. 2671–2675 (2013)
55. D.Z. Du, K.I. Ko, X. Hu, *Design and Analysis of Approximation Algorithms* (Springer, New York, 2012)
56. D. Easley, J. Kleinberg, *Networks, Crowds, and Markets: Reasoning about a Highly Connected World* (Cambridge University Press, Cambridge, 2010)
57. P.W. Eastwick, W.L. Gardner, Is it a game? Evidence for social influence in the virtual world. Soc. Influ. **4**(1), 18–32 (2009)
58. S.M. Elias, A.R. Pratkanis, Teaching social influence: demonstrations and exercises from the discipline of social psychology. Soc. Influ. **1**(2), 147–162 (2006)
59. R. Ennals, D. Byler, J.M. Agosta, B. Rosario, What is disputed on the web? in *Proceedings of the 4th Workshop on Information Credibility (WICOW)* (2010), pp. 67–74
60. factcheck.org, https://www.factcheck.org/
61. M. Faloutsos, P. Faloutsos, C. Faloutsos, On power-law relationships of the internet topology, in *Proceedings of the Annual Conference of the ACM Special Interest Group on Data Communication (SIGCOMM)* (1999), pp. 251–262
62. L. Fan, Z. Lu, W. Wu, B.M. Thuraisingham, H. Ma, Y. Bi, Least cost rumor blocking in social networks, in *ICDCS* (2013), pp. 540–549
63. L. Fan, W. Wu, X. Zhai, K. Xing, W. Lee, D.Z. Du, Maximizing rumor containment in social networks with constrained time. Soc. Netw. Anal. Min. **4**(1), 214 (2014)

64. U. Feige, A threshold of ln n for approximating set cover. J. ACM **45**(4), 634–652 (1998)
65. T.L. Fond, J. Neville, Randomization tests for distinguishing social influence and homophily effects, in *Proceeding of the 19th International Conference on World Wide Web (WWW)* (2010), pp. 601–610
66. N. Fountoulakis, K. Panagiotouy, T. Sauerwaldz, Ultra-fast rumor spreading in social networks, in *Proceedings of the Twenty-third Annual ACM-SIAM Symposium on Discrete Algorithms* (2012), pp. 1642–1660
67. L.A. Fourt, J.W. Woodlock, Early prediction of market success for grocery products. J. Mark. **25**, 31–38 (1960)
68. J.H. Fowler, N.A. Christakis, The dynamic spread of happiness in a large social network: longitudinal analysis over 20 years in the Framingham Heart Study. Br. Med. J. **337**, a2338 (2008)
69. L.C. Freeman, A set of measure of centrality based on betweenness. Sociometry **40**, 35 (1977)
70. L.C. Freeman, Centrality in social networks: conceptual clarification. Soc. Netw. **1**, 215–239 (1979)
71. J. Garrison, C. Knoll, Prop. 8 opponents rally across california to protest gay-marriage ban. Los Angeles Times (2008, November). http://www.latimes.com/local/la-me-prop816-2008nov16-story.html#page=1.html
72. M. Girvan, M.E.J. Newman, Community structure in social and biological networks. Proc. Natl. Acad. Sci. (PNAS) **99**, 7821–7826 (2002)
73. J. Goldenberg, B. Libai, E. Muller, Talk of the network: a complex systems look at the underlying process of word-of-mouth. Mark. Lett. **12**, 211–223 (2001)
74. S. Goyal, M. Kearns, Competitive contagion in networks, in *Proceedings of the 44th Symposium on Theory of Computing* (ACM, New York, 2012), pp. 759–774
75. A. Goyal, F. Bonchi, L.V. Lakshmanan, Learning influence probabilities in social networks, in *Proceedings of the 3rd ACM International Conference on Web Search and Data Mining* (2010), pp. 241–250
76. A. Goyal, W. Lu, L. Lakshmanan, A data-based approach to social influence maximization. *Proc. VLDB Endowment* **5**(1), 73–84 (2011)
77. M. Granovetter, The strength of weak ties. Am. J. Sociol. **78**(6), 1360 (1973)
78. M. Granovetter, Threshold models of collective behavior. Am. J. Sociol. **83**(6), 1420–1443 (1978)
79. M. Granovetter, Economic action and social structure: the problem of embeddedness. Am. J. Sociol. **91**(3), 481–510 (1985)
80. A. Gupta, P. Kumaraguru, C. Castillo, P. Meier, Tweetcred: real-time credibility assessment of content on twitter, in *Proceedings of the International Conference on Social Informatics* (2014), pp. 228–243
81. I. Guyon, J. Weston, S. Barnhill, V. Vapnik, Gene selection for cancer classification using support vector machines. Mach. Learn. **46**(1–3), 389–422 (2002)
82. B. Hajian, T. White, Modelling influence in a social network: Metrics and evaluation, in *IEEE International Conference on Privacy, Security, Risk, and Trust, and IEEE International Conference on Social Computing* (2011), pp. 497–500
83. D. Halliday, R. Resnick, J. Walker, *Fundamentals of Physics*, 8th edn. (Wiley, New Delhi, 2007)
84. X. He, G. Song, W. Chen, Q. Jiang, Influence blocking maximization in social networks under the competitive linear threshold model, in *SDM* (2012)
85. H.W. Hethcote, The mathematics of infectious diseases. SIAM Rev. **42**, 599–653 (2000)
86. K.M. Heussner, Enough already! 7 twitter hoaxes and half-truths. ABC News (2010, January)
87. S. Hill, F. Provost, C. Volinsky, Network-based marketing: identifying likely adopters via consumer networks. Stat. Sci. **21**, 256–276 (2006)
88. P. Holme, M.E.J. Newman, Nonequilibrium phase transition in the coevolution of networks and opinions. Phys. Rev. **74**, 056–108 (2006)
89. A. Java, P. Kolari, T. Finin, T. Oates, Modeling the spread of influence on the blogosphere, in *Proceeding of the 15th International Conference on World Wide Web (WWW)* (2006)

90. J. Jiang, S. Wen, S. Yu, Y. Xiang, W. Zhou, Identifying propagation sources in networks: state-of-the-art and comparative studies. IEEE Commun. Surv. Tutorials **19**, 465–481 (2017)

91. B. Karrer, M. Newman, Competing epidemics on complex networks. Phys. Rev. E **84**(3), 036106 (2011)

92. L. Katz, A new index derived from sociometric data analysis. Psychometrika **18**, 39–43 (1953)

93. D. Kempe, J. Kleinberg, E. Tardos, Maximizing the spread of influence through a social network, in *Proceedings of the 9th ACM SIGKDD Conference on Knowledge Discovery and Data Mining* (2003), pp. 137–146

94. D. Kempe, J. Kleinberg, E. Tardos, Influential nodes in a diffusion model for social networks, in *ICALP* (2005), pp. 1127–1138

95. A. Khrabrov, G. Cybenko, Discovering influence in communication networks using dynamic graph analysis, in *Social Computing/IEEE International Conference on Privacy, Security, Risk and Trust*, vol. 0 (2010), pp. 288–294

96. M. Kimura, K. Saito, Tractable models for information diffusion in social networks, in *Proceedings of the 10th European Conference on Principles and Practice of Knowledge Discovery in Databases* (2006), pp. 259–271

97. M. Kimura, K. Saito, R. Nakano, Extracting influential nodes for information diffusion on a social network, in *AAAI* (AAAI Press, Palo Alto, 2007), pp. 1371–1376

98. M. Kimura, K. Saito, H. Motoda, Minimizing the spread of contamination by blocking links in a network, in *Proceedings of the 23rd AAAI Conference on Artificial Intelligence* (2008)

99. J. Kleinberg, Navigation in a small world. Nature **406**, 845–855 (2000)

100. J. Kleinberg, The small-world phenomenon: An algorithmic perspective, in *Proceedings of the 32nd ACM Symposium on Theory of Computing (STOC)* (2000)

101. J. Kleinberg, S. Lawrence, Authoritative sources in a hyperlinked environment. J. ACM **46**, 604–632 (1999)

102. J. Kleinberg, S. Lawrence, The structure of the web. Science **294**, 1849–1850 (2001)

103. B. Klimt, Y. Yang, Introducing the Enron corpus, in *CEAS* (2004)

104. J. Kostka, Y.A. Oswald, R. Wattenhofer, Word of mouth: Rumor dissemination in social networks, in *SIROCCO* (2008), pp. 185–196

105. F.R. Kschischang, S. Member, B.J. Frey, H. Andrea Loeliger, Factor graphs and the sumproduct algorithm. IEEE Trans. Inf. Theory **47**, 498–519 (2001)

106. R. Kumar, P. Raghavan, S. Rajagopalan, A. Tomkins, Trawling the web for emerging cyber-communities. Comput. Netw. **31**, 1481–1493 (1999)

107. R. Kumar, J. Novak, A. Tomkins, Structure and evolution of online social networks, in *Proceedings of the 12th ACM SIGKDD International Conference on Knowledge Discovery and Data Mining (KDD), Philadelphia, PA* (2006)

108. H. Kwak, C. Lee, H. Park, S. Moon, What is Twitter, a social network or a news media? in *Proceedings of the 19th International Conference on World Wide Web* (2010), pp. 591–600. http://an.kaist.ac.kr/traces/WWW2010.html

109. H. Kwak, C. Lee, H. Park, S. Moon, What is Twitter, a social network or a news media? (2010). http://an.kaist.ac.kr/traces/WWW2010.html

110. A. Lancichinetti, S. Fortunato, Community detection algorithms: a comparative analysis. Phys. Rev. E **80**, 056117 (2009)

111. T. Lappas, E. Terzi, D. Gunopulos, H. Mannila, Finding effectors in social networks, in *KDD* (2010)

112. P. Lazarsfeld, R.K. Merton, Friendship as a social process: a substantive and methodological analysis. Freedom Control Mod. Soc. **18**, 18–66 (1954)

113. J. Leskovec, Stanford large network dataset collection (2009). http://snap.stanford.edu/data/index.html

114. J. Leskovec, J.M. Kleinberg, C. Faloutsos, Graph evolution: densification and shrinking diameters. ACM Trans. Knowl. Discov. Data **1**(1), 2 (2007)

115. J. Leskovec, A. Krause, C. Guestrin, C. Faloutsos, J. Van-Briesen, N. Glance, Cost-effective outbreak detection in networks, in *Proceedings of the 13th ACM SIGKDD Conference on Knowledge Discovery and Data Mining* (2007), pp. 420–429

116. J. Leskovec, L. Backstrom, J. Kleinberg, Meme-tracking and the dynamics of the news cycle, in *Proceedings of the 15th ACM SIGKDD International Conference on Knowledge Discovery and Data Mining (KDD)* (2009), pp. 497–506
117. L. Li, D. Alderson, J.C. Doyle, W. Willinger, Towards a theory of scale-free graphs: definitions, properties, and implications. Internet Math. **2**(4), 431–523 (2006)
118. D. Liben-Nowell, J. Novak, R. Kumar, P. Raghavan, A. Tomkins, Geographic routing in social networks. Proc. Natl. Acad. Sci. (PNAS) **102**(33), 11623–11628 (2005)
119. B. Liu, G. Cong, D. Xu, Y. Zeng, Time constrained influence maximization in social networks, in *IEEE 12th International Conference on Data Mining (ICDM)* (2012), pp. 439–448
120. D. Lopez-Pintado, Diffusion in complex social networks. Games Econ. Behav. **62**(2), 573–590 (2008)
121. T. Lou, J. Tang, J. Hopcroft, Z. Fang, X. Ding, Learning to predict reciprocity and triadic closure in social networks. ACM Trans. Knowl. Discov. Data (TKDD) **7**(2), 5 (2013)
122. Z. Lu, W. Zhang, W. Wu, B. Fu, D.-Z. Du, Approximation and inapproximation for the influence maximization problem in social networks under deterministic linear threshold model, in *ICDCS Workshops* (IEEE Computer Society, Washington DC, 2011), pp. 160–165
123. W. Luo, W.P. Tay, Identifying multiple infection sources in a network, in *Proceedings of the Asilomar Conference on Signals, Systems and Computers* (2012)
124. W. Luo, W.P. Tay, Estimating infection sources in a network with incomplete observations, in *Proceedings of the IEEE Global Conference on Signal and Information Processing (GlobalSIP)* (2013), pp. 301–304
125. W. Luo, W.P. Tay, Finding an infection source under the sis model, in *IEEE International Conference on Acoustics, Speech, and Signal Processing (ICASSP)* (2013)
126. W. Luo, W.P. Tay, M. Leng, Identifying infection sources and regions in large networks. IEEE Trans. Signal Process. **61**, 2850–2865 (2012)
127. S. Maftoon, Quake rumour sends thousands into Ghazni streets (2012). http://www.pajhwok.com/en/2012/08/20/quake-rumour-sends-thousands-ghazni-streets
128. V. Mahajan, E. Muller, F. Bass, New product diffusion models in marketing: a review and directions for research. J. Mark. **54**, 1–26 (1990)
129. E. Mansfield, Technical change and the rate of imitation. Econometrica **29**, 741–766 (1961)
130. M. Mathioudakis, F. Bonchi, C. Castillo, A. Gionis, A. Ukkonen, Sparsification of influence networks, in *Proceedings of the 17th ACM SIGKDD International Conference on Knowledge Discovery and Data Mining* (2011), pp. 529–537
131. S.C. Mednick, N.A. Christakis, J.H. Fowler, The spread of sleep loss influences drug use in adolescent social networks. PLoS One **5**(3), e9775 (2010)
132. M. Mendoza, B. Poblete, C. Castillo, Twitter under crisis: can we trust what we RT? in *Proceedings of the First Workshop on Social Media Analytics (SOMA)* (2010), pp. 71–79
133. P. Metaxas, S. Finn, E. Mustafaraj, Using twittertrails.com to investigate rumor propagation, in *Proceedings of the Eighteenth ACM Conference Companion Computer-Supported Cooperative Work and Social Computing* (2015), pp. 69–72
134. S. Milgram, The small world problem. Psychol. Today **2**, 60 (1967)
135. E. Morozov, Swine flu. Twitter's power to misinform. Foreign Policy (2009)
136. E. Mossel, S. Roch, On the submodularity of influence in social networks, in *Proceedings of the Thirty-ninth Annual ACM Symposium on Theory of Computing (STOC)* (2007), p. 128
137. A. Mouravski, Influence maximization on families of graphs. Thesis, 2011
138. S.A. Myers, J. Leskovec, Clash of the contagions: cooperation and competition in information diffusion, in *2012 IEEE 12th International Conference on Data Mining (ICDM)* (IEEE, Piscataway, 2012), pp. 539–548
139. R. Narayanam, Y. Narahari, Determining the top-k nodes in social networks using the shapley value, in *Proceedings of the 7th International Joint Conference on Autonomous Agents and Multiagent Systems* (2008), pp. 1509–1512
140. F. Nel, M.J. Lesot, P. Capet, T. Delavallade, Rumour detection and monitoring in open source intelligence: understanding publishing behaviours as a prerequisite, in *Proceedings of the Terrorism and New Media Conference* (2010)

141. G.L. Nemhauser, L.A. Wolsey, M. Fisher, An analysis of the approximations for maximizing submodular set functions. Math. Program. **14**, 265–294 (1978)
142. J. Neville, O. Simsek, D. Jensen, Autocorrelation and relational learning: challenges and opportunities, in *Proceedings of the ICML-04 Workshop on Statistical Relational Learning* (2004)
143. M.E.J. Newman, The structure of scientific collaboration networks. Proc. Natl. Acad. Sci. (PNAS) **98**, 409–415 (2001)
144. M.E.J. Newman, A measure of betweenness centrality based on random walks. Soc. Netw. **27**(1), 39–54 (2005)
145. M.E.J. Newman, *Networks: An Introduction.* (Oxford University Press, Oxford, 2010)
146. M. Newman, Network data (2013). http://www-personal.umich.edu/~mejn/netdata/
147. N.P. Nguyen, G. Yan, M.T. Thai, S. Eidenbenz, Containment of misinformation spread in online social networks, in *Proceedings of the 4th Annual ACM Web Science Conference (WebSci)* (2012), pp. 213–222
148. D.T. Nguyen, N.P. Nguyen, M.T. Thai, Sources of misinformation in online social networks: who to suspect? in *Proceedings of the IEEE Military Communications Conference MILCOM* (2012), pp. 1–6
149. H.T. Nguyen, M.T. Thai, T.N. Dinh, Stop-and-stare: optimal sampling algorithms for viral marketing in billion-scale networks, in *Proceedings of the 2016 International Conference on Management of Data* (2016), pp. 695–710
150. L. Page, S. Brin, R. Motwani, T. Winograd, The pagerank citation ranking: bringing order to the web. Technical report, Stanford University (1998)
151. N. Pathak, A. Banerjee, J. Srivastava, A generalized linear threshold model for multiple cascades, in *2010 IEEE 10th International Conference on Data Mining (ICDM)* (IEEE, Piscataway, 2010), pp. 965–970
152. W. Peterson, N. Gist, Rumor and public opinion. Am. J. Sociol. **57**, 159–167 (1951)
153. A.G. Phadke, J.S. Thorp, *Computer Relaying for Power Systems* (Wiley, Chichester, 1988)
154. P.C. Pinto, P. Thiran, M. Vetterli, Locating the source of diffusion in large-scale networks. Phys. Rev. Lett. **109**, 068–702 (2012)
155. I. Pool, M. Kochen, Contacts and influence. Soc. Netw. **1**, 1–48 (1978)
156. B.A. Prakash, A. Beutel, R. Rosenfeld, C. Faloutsos, Winner takes all: competing viruses or ideas on fair-play networks, in *Proceedings of the 21st International Conference on World Wide Web* (ACM, New York, 2012), pp. 1037–1046
157. B.A. Prakash, J. Vreeken, C. Faloutsos, Spotting culprits in epidemics: how many and which ones? in *IEEE International Conference on Data Mining (ICDM)* (2012), pp. 11–20
158. V. Qazvinian, E. Rosengren, D.R. Radev, Q. Mei, Rumor has it: identifying misinformation in microblogs, in *Proceedings of the 2011 Conference on Empirical Methods in Natural Language Processing* (2011), pp. 1589–1599
159. A. Rad, M. Benyoucef, Towards detecting influential users in social networks, in *E-Technologies: Transformation in a Connected World: 5th International Conference, MCETECH, Les Diablerets*, vol. 78 (2011)
160. J. Ratkiewicz, M. Conover, M. Meiss, B. Goncalves, S. Patil, A. Flammini, F. Menczer, Truthy: mapping the spread of astroturf in microblog streams, in *Proceedings of the 20th International Conference Companion on WWW* (2011), pp. 249–252
161. M. Richardson, P. Domingos, Mining knowledge-sharing sites for viral marketing, in *Proceedings of the Eighth ACM SIGKDD International Conference on Knowledge Discovery and Data Mining*, pp. 61–70 (ACM, New York, 2002)
162. M.G. Rodriguez, D. Balduzzi, B. Schölkopf, Uncovering the temporal dynamics of diffusion networks, in *ICML* (2011), pp. 561–568
163. J.N. Rosenquist, J. Murabito, J.H. Fowler, N.A. Christakis, The spread of alcohol consumption behavior in a large social network. Ann. Internal Med. **152**(7), 426–W141 (2010)
164. K. Saito, R. Nakano, M. Kimura, Prediction of information diffusion probabilities for independent cascade model, in *Knowledge-Based Intelligent Information and Engineering Systems* (Springer, 2008), pp. 67–75

165. K. Saito, K. Ohara, Y. Yamagishi, M. Kimura, H. Motoda, Learning diffusion probability based on node attributes in social networks, in *ISMIS* (2011), pp. 153–162
166. T. Sakaki, M. Okazaki, Y. Matsuo, Earthquake shakes twitter users: real-time event detection by social sensors, in *Proceedings of the 19th International Conference on World Wide Web* (2010)
167. P. Sarkar, A. Moore, Dynamic social network analysis using latent space models, in *ACM SIGKDD Explorations Newsletter*, vol. 7(2) (2005), pp. 31–40
168. J. Schumpeter, U. Bakhays, *The Theory of Economics Development* (Springer, New York, 2003)
169. J. Scripps, P.N. Tan, A.H. Esfahanian, Measuring the effects of preprocessing decisions and network forces in dynamic network analysis, in *Proceedings of the 15th ACM SIGKDD International Conference on Knowledge Discovery and Data Mining (KDD)* (2009), pp. 747–756
170. E. Seo, P. Mohapatra, T. Abdelzaher, Identifying rumors and their sources in social networks, in *SPIE Defense, Security, and Sensing* (2012)
171. D. Shah, T. Zaman, Finding sources of computer viruses in networks: theory and experiment. *Proceedings of the ACM Sigmetrics*, vol. 15 (2010), pp. 5249–5262
172. D. Shah, T. Zaman, Rumors in a network: who's the culprit? IEEE Trans. Inf. Theory **57**(8), 5163–5181 (2011)
173. D. Shah, T. Zaman, Rumor centrality: a universal source detector, in *SIGMETRICS* (2012), pp. 199–210
174. C. Shao, G. Ciampaglia, A. Flammini, F. Menczerr, Hoaxy: a platform for tracking online misinformation, in *Proceedings of the Twenty-fifth International Conference Companion World Wide Web* (2016), pp. 745–750
175. S. Shelke, V. Attar, Source detection of rumor in social network a review. Online Soc. Netw. Media **9**, 30–42 (2019)
176. X. Shi, J. Zhu, R. Cai, L. Zhang, User grouping behavior in online forums, in *Proceedings of the 15th ACM SIGKDD International Conference on Knowledge Discovery and Data Mining (KDD)* (2009), pp. 777–785
177. L. Shu, M. Mukherjee, X. Xu, K. Wang, X. Wu, A survey on gas leakage source detection and boundary tracking with wireless sensor networks. IEEE Access **4**, 1700–1715 (2016)
178. G. Siganos, S.L. Tauro, M. Faloutsos, Jellyfish: a conceptual model for the as internet topology. J. Commun. Netw. **8**(3), 339–350 (2006)
179. P. Singla, M. Richardson, Yes, there is a correlation: from social networks to personal behavior on the web, in *Proceeding of the 17th International Conference on World Wide Web (WWW)* (2008), pp. 655–664
180. P. Smolensky, Information processing in dynamical systems: Foundations of harmony theory, in *Parallel Distributed Processing*, ed. by D. E. Rumelhart, J. L. McClelland, vol. 1, (MIT Press, Cambridge, 1986), pp. 194–281
181. Snopes.com, https://www.snopes.com/
182. J. Sun, J. Tang, A survey of models and algorithms for social influence analysis, in *Social Network Data Analytics* (Springer, Boston, 2011), pp. 177–214
183. L. Sun, W. Huang, P.S. Yu, W. Chen, Multi-round influence maximization, in *Proceedings of the 24th ACM SIGKDD International Conference on Knowledge Discovery and Data Mining* (2018), pp. 2249–2258
184. L. Tang, H. Liu, Relational learning via latent social dimensions, in *Proceedings of the 15th ACM SIGKDD International Conference on Knowledge Discovery and Data Mining (KDD)* (2009), pp. 817–826
185. L. Tang, H. Liu, Scalable learning of collective behavior based on sparse social dimensions, in *Proceeding of the 18th ACM Conference on Information and Knowledge Management (CIKM)* (2009), pp. 1107–1116
186. J. Tang, J. Sun, C. Wang, Z. Yang, Social influence analysis in large-scale networks, in *Proceedings of the 15th ACM SIGKDD International Conference on Knowledge Discovery and Data Mining* (2009), pp. 807–816

187. Y. Tang, X. Xiao, Y. Shi, Influence maximization: near-optimal time complexity meets practical efficiency, in *Proceedings of SIGMOD International Conference on Management of Data* (2014), pp. 75–86

188. Y. Tang, Y. Shi, X. Xiao, Influence maximization in near-linear time: a martingale approach, in *Proceedings of the 2015 ACM SIGMOD International Conference on Management of Data* (2015), pp. 1539–1554

189. R.T.R. Tracker, http://www.emergent.info/

190. D. Trpevski, W.K.S. Tang, L. Kocarev, Model for rumor spreading over networks. Phys. Rev. E **81**, 056102 (2010)

191. J. Tsai, T.H. Nguyen, M. Tambe, Security games for controlling contagion, in *AAAI*, ed. by J. Hoffmann, B. Selman (AAAI Press, Palo Alto, 2012)

192. J. Tsai, Y. Qian, Y. Vorobeychik, C. Kiekintveld, M. Tambe, Bayesian security games for controlling contagion, in *SocialCom* (IEEE, Piscataway, 2013), pp. 33–38

193. C.E. Tsourakakis, Fast counting of triangles in large real networks without counting: Algorithms and laws, in *Eighth IEEE International Conference on Data Mining*, (2008), pp. 608–617

194. Twitter, https://about.twitter.com

195. V. Tzoumas, C. Amanatidis, E. Markakis, A game-theoretic analysis of a competitive diffusion process over social networks, in *WINE*, ed. by P.W. Goldberg (2012), pp. 1–14

196. J. Ugandera, L. Backstromb, C. Marlowb, J. Kleinberg, Structural diversity in social contagion. Proc. Natl. Acad. Sci. **109**(20), 7591–7592 (2012)

197. Y. Wang, G. Cong, G. Song, K. Xie, Community-based greedy algorithm for mining top-k influential nodes in mobile social networks, in *KDD* (2010), pp. 1039–1048

198. C. Wang, W. Chen, Y. Wang, Scalable influence maximization for independent cascade model in large-scale social networks. Data Min. Knowl. Discov. **25**(3), 545–576 (2012)

199. S. Wasserman, K. Faust, *Social Networks Analysis: Methods and Applications* (Cambridge University Press, Cambridge, 1994)

200. D.J. Watts, S.H. Strogatz, Collective dynamics of small-world networks. Nature **393**, 440–442 (1998)

201. M. Welling, G.E. Hinton, A new learning algorithm for mean field boltzmann machines, in *Proceedings of International Conference on Artificial Neural Network (ICANN)* (2001), pp. 351–357

202. D. Westermana, P.R. Spenceb, B.V.D. Heide, A social network as information: the effect of system generated reports of connectedness on credibility on twitter. Comput. Hum. Behav. **28**(1), 199–206 (2012)

203. C. Wildeman, A.V. Papachristos, Network exposure and homicide victimization in an African American community. Am. J. Public Health **337**, 337 (2013)

204. X. Wu, X. Zhu, G.Q. Wu, W. Ding, Data mining with big data. IEEE Trans. Knowl. Data Eng. **26**(1), 97–107 (2014)

205. R. Xiang, J. Neville, M. Rogati, Modeling relationship strength in online social networks, in *Proceeding of the 19th International Conference on World Wide Web (WWW)* (2010), pp. 981–990

206. W. Xu, H. Chen, Scalable rumor source detection under independent cascade model in online social networks, in *Proceedings of the 2015 Eleventh International Conference on Mobile Ad-hoc and Sensor Networks (MSN)* (2015), pp. 236–242

207. W. Xu, Z. Lu, W. Wu, Z. Chen, A novel approach to online social influence maximization. J. Soc. Netw. Anal. Min. (SNAM) **4**(1), 153–164 (2014)

208. Yahoo News, Twitter rumor leads to sharp increase in the price of oil (2012). http://news.yahoo.com/blogs/technology-blog/twitter-rumor-leads-sharp-increase-price-oil-173027289.html

209. Q. Yan, S. Guo, D. Yang, Influence maximizing and local influenced community detection based on multiple spread model, in *ADMA, Part II*. LNAI 7121 (2011), pp. 82–95

210. Z. Zhang, W. Xu, W. Wu, D.-Z. Du, A novel approach for detecting multiple rumor sources in networks with partial observations. J. Combinat. Optim. **33**, 1–15 (2015)
211. K. Zhu, L. Ying, Information source detection in the sir model: a sample path based approach, in *Information Theory and Applications Workshop (ITA)*, vol. 24(1) (2013), pp. 408–421
212. K. Zhu, L. Ying, A robust information source estimator with sparse observations. *IEEE International Conference on Computer Communications (INFOCOM)*, vol. 1 (2014), p. 3

Printed in the United States
By Bookmasters